Concrete Systems for Homes and Low-Rise Construction

Concrete Systems for Homes and Low-Rise Construction

A Portland Cement Association Guide

Pieter A. VanderWerf
Ivan S. Panushev
Mark Nicholson
Daniel Kokonowski

McGraw-Hill

New York   Chicago   San Francisco   Lisbon   London   Madrid
Mexico City   Milan   New Delhi   San Juan   Seoul
Singapore   Sydney   Toronto

The McGraw·Hill Companies

Cataloging-in-Publication Data is on file with the Library of Congress.

1 2 3 4 5 6 7 8 9 0 DOC/DOC 0 1 0 9 8 7 6 5

ISBN 0-07-145236-2

The sponsoring editor for this book was Larry S. Hager and the production supervisor was Richard Ruzycka. It was set in Garamond by Matrix Publishing Services. The art director for the cover was Anthony Landi.

Printed and bound by RR Donnelley.

 This book was printed on recycled, acid-free paper containing a minimum of 50% recycled, de-inked fiber.

McGraw-Hill books are available at special quantity discounts to use as premiums and sales promotions, or for use in corporate training programs. For more information, please write to the Director of Special Sales, McGraw-Hill Professional, Two Penn Plaza, New York, NY 10121-2298. Or contact your local bookstore.

Contents

Acknowledgments

It took a great deal of help to compile so much information on the wide range of construction products covered in this book. It is impossible to thank all individuals and organizations that provided information, contributed photographs, or proofed parts of the text. However, the authors wish to recognize some who made important contributions. They are listed here:

Aerated Autoclaved Concrete Products Association, www.aacpa.org
Aerated Concrete Corp. of America, www.accoaac.com
AERCON Florida, LLC, www.aerconfl.com
Aercon Industries, LLC, www.aerconind.com
American ConForm Industries Inc., www.smartblock.com
American PolySteel, LLC, www.polysteel.com
Architectural/Residential Technologies, Inc., www.arit.com
Arxx Building Products, www.arxxbuild.com
Avantec, Ltd., www.avantecltd.com
Azar Mortarless Building Systems, Inc., www.azarblock.com
Becica and Associates, www.becica.com
Besser Company, www.besser.com
BlockJoist Co., LLC, www.blockjoist.com
Bridgeworks, Inc., www.bridgeworks.com
Brock Savage Construction, Georgia
Canam Group, Inc., www.hambrosystems.com
Cement Association of Canada, www.cement.ca
CEMEX USA, www.cemexusa.com
Central Concrete, www.cencrete.com
CertainTeed Corporation, www.certainteed.com
Chameleon Cast Wall System, LLC, www.chameleonwall.com
Cheng Concrete Exchange, www.concretexchange.com
Composite Technologies Corporation, www.thermomass.com
Concrete Construction magazine, www.concreteconstruction.net.
Concrete Foundations Association of North America, www.cfawalls.org
Concrete Home Building Council, www.nahb.org
Concrete Homes Council, www.concretehomescouncil.com
ConcreteNetwork.com, www.ConcreteNetwork.com
ConForm Pacific, Inc., www.smartblock.com

CON/STEEL Tilt-Up Systems, www.consteel.com
Construction Technology Laboratories, www.ctlgroup.com
Continental Cast Stone Manufacturing, Inc., www.caststone.net
Degussa, Inc., www.degussa.com
Distinctive Concrete Boston, Inc., www.distinctiveconcrete.com
Dukane Precast, Inc., www.dukaneprecast.com
ECO-BLOCK, LLC, www.eco-block.com
E-Crete, www.e-crete.com
Efficient Wall Systems of Florida, LLC, www.e-wall.info
Eggert Construction, LLC, Branford, Connecticut
Eldorado Stone Operations, LLC, www.eldoradostone.com
Elite Custom Builders, Westerley, RI
Engelman Construction, Macungie, PA
E.P. Henry Company, www.ephenry.com
Floor Seasons, Inc., www.floorseasons.com
Florida Concrete Products Association, www.floridaconcrete.org
Fritz Concrete, Chepstow, ON
G. Mustapick Companies, Florida
Grace Construction Products, www.na.graceconstruction.com
Hambro Structural Systems, www.hambrosystems.com
Hanley Wood, LLC, www.hanley-wood.com
Hadrian Tridi-Systems, www.tridipanel.com
Hedrick Brother's Construction, www.hedrickbrothers.com
Hughes Construction, North Carolina
ICF Accessories, Longville, MN
ICF Builders, www.icfbuilders.com
Innovative Brick, www.Mbrick.com
Insulating Concrete Form Association, www.forms.org
INSUL-DECK, LLC, www.insul-deck.org
Interlocking Concrete Paver Institute, www.icpi.org
James Hardie Building Products, www.jameshardie.com
Kaw Valley Habitat for Humanity, Kansas City, MO
Kings Material, Inc., www.kingsmaterial.com
Korte Company, www.korteco.com
LaBarge Engineering and Contracting, Harwich, MA
Lafarge North America, www.lafarge.com
Landscape Development, Inc., www.landscapedevelopmentinc.com
Ledbetter Masonry & Construction, Inc., Florida
Lite-Form Technologies, www.liteform.com
Lithko Corp., www.lithko.com
L.M. Scofield Company, www.scofield.com
Logix Insulated Concrete Forms, Ltd., www.logixicf.com
Manning Building Supplies/Quick Wall, www.mbs-corp.com
Metal Lath and Stucco Co., Alabama

Meyer Brothers Building Co., www.meyerbro.com
Midwest Block and Brick, www.midwestblock.com
Monier Lifetile, LLC, www.monierlifetile.com
Moore Masonry, www.mooremasonry.com
National Concrete Masonry Association, www.ncma.org
National Precast Concrete Association, www.precast.org
National Ready-Mix Concrete Association, www.nrmca.org
New England Concrete Masonry Association, www.necma.com
NUDURA Corporation, www.nudura.com
Nu Walls, Inc., www.nuwalls.com
Oke Woodsmith, www.okewoodsmith.com
Oldcastle® Building Systems Group, www.oldcastle.com
OldCastle Precast®, www.oldcastle-precast.com
Owens Corning, www.owenscorning.com
Pavers by Ideal, www.idealconcreteblock.com
Perform Wall, LLC, www.performwall.com
Phil-Insul Corporation, www.integraspec.com
Phoenix Systems & Components, Inc., www.phoenixicf.com
P.K. Construction, Lincoln, NE
Polysteel Southeast Distributors, www.polysteelsoutheast.com
The Portland Cement Association, www.cement.org
Precast/Prestressed Concrete Institute, www.pci.org
Precise Forms, Inc., www.preciseforms.com
Quad-Lock Building Systems, Ltd., www.quadlock.com
Ram Builders, www.rambuilders.com
Randy Adams Construction, www.randy-adams.com
Reward Wall Systems, Inc., www.rewardwalls.com
Royal Building Technologies, www.rbsdirect.com
Royall Wall Systems, Inc., www.royallwall.com
Seadore Masonry, Inc., www.seadoremasonry.com
Seretta Construction, www.seretta.com
SI Corporation, www.sind.com
Southeast Florida Polysteel, Florida
South River Construction, www.southriverconstruction.com
The Spancrete Group, Inc., www.spancrete.com
Spaulding Brick Company, www.spauldingbrick.com
Speedfloor Holdings, Ltd., www.arit.com
Stampcrete International, Ltd., www.stampcrete.com
Steinbicker & Associates, Inc., www.sai-engineers.com
Stone Soup Concrete, www.stonesoupconcrete.com
Strata Systems, Inc., www.geogrid.com
Superior Walls, www.superiorwalls.com
Suspended Concrete Systems, Ltd., www.suspendedconcrete.co.nz
Tadros Associates, www.tadrosassociates.com

Tejas Textured Stone, www.tejasstone.com
Tile Roofing Institute, www.tileroofing.org
Tilt-up Concrete Association, www.tilt-up.org
Tiltwall Systems, Georgia
US Paverscape, www.uspaverscape.com
Vinyl Technologies, Inc., www.vbuck.com
Wall-Ties and Forms, Inc., www.wallties.com
Walker Engineering, Birmingham, AL
Westbrook Concrete Block Co., Inc., www.westbrookblock.com
Western Forms, Inc., www.westernforms.com
Westile Roofing Products, www.westile.com
Wisconsin Thermo-Form, Inc., www.tfsystem.com
Woodland Construction Company, www.woodlandconstruction.com

Introduction

When we released the book *Concrete Homebuilding Systems* ten years ago, it was clear we hit a nerve. Although no one realized it at the time, contractors for low-rise buildings were beginning a rush to concrete- and cement-based products. Yet there was no central source to acquaint them with these. The demand for information was great, but there were few resources to supply it. Our book was one of the first, and it sold far beyond projections.

Recently the Portland Cement Association decided that it was time for an update. The use of concrete products in small buildings has grown rapidly, and contactors' need for information has grown with it.

The scope of this book is much broader than the last one. It covers all major concrete and cement-based products, not just wall systems. It adds concrete floor and roof systems, exterior finish products, landscape products, and decorative applications. It also covers the use of these products in all small, low-rise buildings, not just single-family homes. That rounds out the focus to include commercial and multifamily buildings up to a size of roughly 20,000 square feet of floor area and 40 feet in height.

This book should be useful for many building professionals, but it is directed first and foremost at the general contractor (GC) who constructs the buildings. We have tried to provide the information the GC needs to decide which of these products make sense for his or her business. The book also helps the GC to get started with the products and get rolling on them with maximum efficiency and minimum fuss.

The organization of the book is intended to be self-explanatory. Part 1 covers general information that applies to many or all the products in the book, such as principles of concrete and ways to measure product costs. Parts 2 through 6 go into specifics on the different categories of products. For example, Part 2 covers wall systems.

Each one of the Parts covering one category of products begins with a section titled "Background." These cover general principles that are important for multiple products in the category. For example, the Background of Part 2 discusses such things as moisture control, sound transmission, and other issues important to understanding walls. After this are the sections on each specific product or system within the category, such as concrete block walls, insulating concrete form walls, precast walls, walls cast with removable forms, and so on. At the end of each Part is a section titled "Developments." This describes things that promise to become important in the future, including new or specialty products, inventions, and trends.

A reasonable plan of attack is to start by reading all of Part 1 and the "Backgrounds" of Parts 2 through 6. This should help decide what specific products will be of most interest and provide the background information to read the individual product sections.

But there is no wrong way to read the book. This is a huge field with tremendous promise for the creators of small buildings. Read, wander, and enjoy.

Other sources of information

There are plenty of good periodicals and Web sites that also provide useful information on the concrete products and systems covered in this book. They are especially useful for current developments and information that changes frequently over time, such as costs and prices. Many of these information sources deal with one or two specific products. They are described in the chapter dealing with the relevant product. But others cover a wide range of concrete products and systems for small buildings. It is worth mentioning them here.

Periodicals

Concrete Construction covers a wide range of topics of interest to the concrete professional. Most of these apply to small buildings. For information go to www.concreteconstruction.net.

Concrete International is the magazine of the American Concrete Institute (ACI). It covers a broad range of topics related to concrete, many of which are relevant to small buildings. Go to www.concreteinternational.com for more information.

Concrete Homes magazine discusses all sorts of concrete products and systems and how they are used in single-family and small multifamily housing. For information, go to www.concretehomesmagazine.com.

Masonry Construction covers all types of masonry products, including concrete block, concrete brick, cast stone, concrete pavers, and segmental retaining walls. These are widely used in small buildings. Get more information about it at www.hanleywood.com.

Permanent Buildings and Foundations covers all sorts of products and systems used in homes and small commercial buildings. Information is available at www.pbf.org.

Residential Concrete is a magazine focused on all sorts of concrete products, uses, and systems in houses. Most of them apply to small nonresidential buildings as well. For information, go to www.hanleywood.com.

Web sites

The independent site www.concretenetwork.com provides information and links related to a wide variety of concrete products and systems used in small buildings.

The Portland Cement Association's site www.concretehomes.com covers every significant product or system used in houses. Most of the information is also relevant to use in other small buildings.

Associations

There are many trade associations that support and promote the use of concrete construction products. The relevant associations are covered in each of the individual product chapters in this book. However, two cover a wide range of products and bear mentioning here.

The Portland Cement Association provides information and technical assistance for virtually all products that contain significant amounts of cement. They provide extensive materials on many products and maintain a professional staff that can answer questions and provide assistance on a wide range of issues. The Association is accessible through its Web site, www.cement.org. As already discussed, it also maintains www.concretehomes.com, an information Web site on concrete products suitable for homes and other small buildings.

A very important new organization in this area is the Concrete Home Building Council (CHBC) of the National Association of Home Builders (NAHB). The NAHB represents the general contractors who construct homes in the United States. The CHBC is a newly-formed arm of the NAHB to support home builders and the leadership of the NAHB in their work with concrete products. They are quickly becoming a key clearinghouse for concrete products information and assistance to the construction community. The CHBC is accessible through the NAHB Web site (www.nahb.org) or by calling the NAHB (202-822-0200).

Concrete Systems for Homes and Low-Rise Construction

PART I

GENERAL

1

Why concrete?

The use of concrete and cement based building products in low-rise construction has risen sharply in the last 20 years. Concrete walls have long dominated foundations, but above-grade they have grown from an estimated 3 percent of single-family homes in 1994 to 16 percent in 2004. They have jumped in commercial construction as well. A few years behind the growth in walls has come a similar growth in concrete floors and roofs.

Interior decorative concrete floors and countertops started with fewer than ten contractors in California in the early 1980s and have grown to thousands in 2005.

Stucco remains one of the most popular of all exterior wall finishes. Fiber-cement siding has grown from nothing to about 10 percent of all lapped siding in North America over the last 25 years. Concrete roof tile has grown from nearly zero to several percent of the coverings for pitched roofs in the same period.

Among landscape products, segmental retaining walls were nonexistent in 1980. Now they make up about half of the retaining walls surrounding new low-rise buildings. Concrete flatwork came seriously into use on the lots of small buildings after World War II, and now it is estimated to make up over 50 percent of the drives and walkways. Concrete pavers were under 1 percent of the paving, and now they are believed to be over 10 percent.

The growth in these materials is projected to continue.

Why has this happened? The answer is that there have been important changes in both supply and demand. On the supply side, there have been dramatic improvements in concrete, concrete products, and the systems that use them. These have made the concrete options more economical, easier to use, higher performing, and more aesthetic. On the demand side, buyers have rising incomes, and they are using some of that money to buy better buildings. They are increasingly looking for materials that are disaster resistant, conducive to comfort, energy efficient, durable, low maintenance, and distinctive looking. All of these are inherent properties of concrete. So the market is shifting to the sort of product that concrete has been all along.

Supply

Advances in concrete technology and product design are discussed throughout this book. Some developments in concrete chemistry give the material new properties. Others reduce concrete's cost without compromising its desirable properties. There are now exceptionally high-strength concretes, lightweight concretes filled with air bubbles, and concretes containing wood fibers for flexure and ease of cutting and nailing.

The possible aesthetics of concrete have grown dramatically. Pigments permit concrete to take any color of the rainbow. Concrete stains provide subtle gradations in color. Surface treatments, such as stamping, fracturing, sandblasting, retarding, tumbling, and grinding, create a wide range of finish textures that give concrete the look and feel of natural materials like stone or novel and high-tech looks not available on any other construction material (Figure 1-1).

1-1 *Applying premium finishes to concrete.* Floor Seasons, Inc.

New accessory products have reduced the cost and increased the variety of shapes and products that can be created out of concrete. Examples are form-work made of foam, steel forms optimized for above-grade walls, steel joists that bond with concrete flatwork to create a floor with composite action, rubber molds that give concrete the exact shape and texture of natural stone, and machines designed to mold small objects like brick, tiles, and paving stones efficiently.

Advances in concrete formulas have made it possible to use more recycled material. This helps reduce cost and makes concrete a more environmentally friendly material. Concrete suppliers have learned to incorporate waste products from other industrial processes. These include fly ash from the coal burned in power plants and slag from such processes as iron and steel production. Concrete masonry and precasting plants now regularly grind their waste material and put it back into the mix instead of discarding it.

During all this time the price of basic concrete has remained relatively stable. In contrast, some other materials have experienced both price increases and declines in availability. The stands of large trees that are easily harvested in North America are dwindling. As a result, the price of dimensional lumber over the last fifteen years has been volatile and has risen more rapidly than prices in general. To get every possible piece of material out of the available trees, suppliers are providing more and more pieces that are twisted, bowed, or shy of stated dimensions. An alternative from the forest products industry is engineered lumber. Engineered lumber consists of dimensional material that is created by gluing a set of smaller pieces together. The result is a quality product, but it can be as much as twice as expensive as conventional lumber.

Steel and asphalt have also experienced greater price volatility and steeper long-term price trends than concrete and concrete products. Their future is difficult to predict, but the factors driving their price fluctuations in the past may continue.

Demand

As buyer demand and incomes rise, increasingly people want the premium benefits of concrete.

Even before any of the improvements in concrete listed here, the material had valuable properties that come "standard" with concrete. For one, any given volume of concrete is inexpensive compared to most other building materials. Concrete is also high in compressive strength, durable, and low in maintenance requirements. With proper steel reinforcing, it has proven to be one of the most resistant materials to such disasters as wind, earthquake, flood, fire, and impact.

Its durability stems from its high resistance to water, moisture, rot, rust, mold, mildew, heat, cold, fire, chemicals, and ultraviolet rays. This durability also leads to a low need for maintenance. The normal forces of wear have little effect on concrete. Treatments to maintain it or repair it are necessary less often than with many other building materials. This is particularly true when the materials are exposed to the elements.

These properties have always been valued by building owners. However, decades ago concrete was rarely chosen for many parts of the small building. In large buildings, it was often used because large structures require great strength, and their surfaces require durability to withstand heavy use. It was used in small buildings for foundations because strength and resistance to water are critical in exposed, below-grade applications. But in every other part of the small building, it was used either never or much less often than alternative materials.

The primary reason for this situation was that the concrete products were often more expensive. Working on tight budgets, designers and owners would lean toward other materials to shave costs. They might want the benefits concrete had to offer, but felt they could not afford them.

But budgets change. Incomes and expectations are slowly but surely rising for everyone: consumers, corporations, and governments. With more money, buyers gradually turn to spending more on all aspects of their lives. That includes their buildings. Eventually they begin to devote more money to get superior building materials.

It is, therefore, not surprising to see people request concrete products more and more often. They have always valued the sorts of properties that concrete possesses. Every year they have more means to pay for them.

Other events have magnified this effect by heightening demand further. Perhaps the clearest of these was Hurricane Andrew in 1992 and the amazing string of four hurricanes that lashed the Southeast in 2004. In the news reports that followed, people saw revealing photographs of the devastation. In many, a light frame building was little more than twisted wreckage, while alongside were reinforced concrete structures that were still largely intact. The resistance of reinforced concrete to wind was not lost on the public. These events served as a strong reminder of the general strength and durability properties of concrete. They led many buyers to consider it seriously for upcoming projects. The Southwestern brush fires of the last ten years had a similar effect on public opinion.

For all of these reasons, people are requesting concrete in growing numbers.

The implications for contractors

The twin forces of supply and demand are affecting the construction of small buildings, like the blades of a scissors cutting through cloth. As incomes rise, more and more buyers opt for the advantages that concrete has always offered. At the same time, there are sharp improvements in the material and the products constructed with it. The improvements have made concrete even more attractive. As these two forces work, the sales of concrete products and systems have naturally risen.

These shifts have left contractors with a need to learn more about concrete products. Customers are requesting them. Traditional contractors lose some jobs because they are unable to provide the concrete products or are unfamiliar with them. Not surprisingly, many contractors confronted by these buyer requests are adopting more concrete.

But some contractors are working concrete products into their construction practices even without any special requests from buyers. Some concrete products offer logistic or cost-saving advantages in construction. Some contractors have decided that they are worth adopting just to get these construction advantages. Still other contractors have decided to offer some of these new products to attract new customers and show the public that their construction is progressive and high quality.

Eventually, almost every contractor will be faced with a decision. As the use of concrete products grows, every contractor will find that more and more of the competition is offering them. They will need to decide whether to adopt them, when, and how. Those who start thinking about it early will have the chance to gain an edge on the competition.

This book is designed to help the contractor evaluate the many promising concrete products for small buildings. It goes into important details and covers all important issues to help the reader understand each product in a short space. It also provides information on how to do further research and go about adopting any product that proves to be of interest.

The remainder of Part 1 provides an overview of the concrete products available. It also covers general principles of concrete and concrete products that are important to the more detailed product descriptions later in the book. The later Parts (2–6) describe the products in depth. Each one covers a class of products or systems, such as wall systems, floor and roof systems, exterior finish products, and so on. Each of these parts begins with more background, followed by one chapter of detailed description for each major product. At the end of each part is a "Developments" chapter that alerts the reader to important specialized products and major trends in the field.

2

What's available?

There are now concrete products for use throughout small buildings. We have divided them into five categories:

1. Exterior walls
2. Floors and roofs
3. Exterior finishes
4. Landscape products
5. Decorative concrete

In each of these categories there are multiple different concrete products available for use. One of the reasons for this is that there are several methods of turning wet concrete into solid products that have been developed over the years. In many cases more than one different method can be used to make a product for the same general purpose.

Methods of fabrication

The major distinct methods of producing concrete products include:

- Cast-in-place with removable forms
- Insulating stay-in-place forms
- Masonry
- Precast
- Tilt-up
- Autoclaved, aerated concrete
- Fiber cement
- Troweling

Each of these methods will produce somewhat different products with different advantages and disadvantages. Some are more useful and practical for certain applications than they are for others. But for many uses, more than one type of product could be used, each with its own distinctive strengths.

Cast-in-place products are created by placing fluid concrete in forms to give the concrete its final shape, at its final location. It does not need to be moved. *Removable forms* are designed to be taken off the concrete after it becomes solid. They may generally be reused for the same purpose later. *Stay-in-place forms* are designed to remain in position permanently. They provide a sort of "skin" for the concrete that may have desirable properties. The *insulating concrete form* (ICF) is a type of stay-in-place form coming into wide use recently. It is made of an insulating material, and remains to provide insulation to the structure (Figure 2-1 and 2-2).

Masonry consists of a set of individual pieces of concrete that are set on one another at the site to make a larger assembly, such as a wall. Concrete masonry products are sometimes divided into two further subgroups: *dry-cast* and *wet-cast*. Dry-cast products are molded with a minimum of water in the concrete mix. Since the relatively "dry" concrete does not flow readily, the molds are vibrated with special machinery to fill them. The process is relatively rapid, and the machines are generally designed to create large numbers of identical or similar pieces of masonry. Wet-cast products are created by placing wet concrete into molds and waiting for it to cure adequately before removing the masonry. This process is more commonly used for concrete units that are made in small quantities or differ from one another (Figure 2-3).

Precast products are also cast wet in a plant. However, they are generally larger pieces for such applications as walls and floors. Most of them are cast on a flat "table" with special low forms arranged as needed to make the edges of the precast units. These pieces are then shipped to the job site and set in place, usually with a crane (Figure 2-4).

Tilt-up concrete is cast at the job site on the ground and then lifted into place. It is used almost exclusively for walls. The wall panels are cast on a flat surface, usually a floor slab. The edges are also defined by low forms (Figure 2-5).

Autoclaved, aerated concrete (AAC) is a form of concrete that includes a large number of tiny air bubbles. This makes it relatively light, a bit softer and easier to cut, and gives it insulating properties. Autoclaved, aerated concrete is manufactured in factories in large sections. These are cut into blocks or panels that are assembled into buildings, in much the same way as conventional concrete masonry units and precast wall panels (Figure 2-6).

Fiber cement is another special material. It is made by combining cement with wood fibers, without the usual sand and stones. This makes it more flexible and easier to cut and drive a nail through. It can be molded or pressed into various shapes to make useful products.

Wet concrete can also be *troweled* in place at the job site. Pressing concrete against a surface with a trowel is useful for building up a finish layer of material.

2-1 *Removable forms for constructing concrete walls.*

2-2 *Insulating concrete forms for constructing concrete walls.* ICF Builders

2-3 *Machinery for molding concrete masonry in a dry-cast process.* Besser Company

2-4 *Operations of a precasting plant.*

2-5 *Formwork for tilt-up wall panels.*

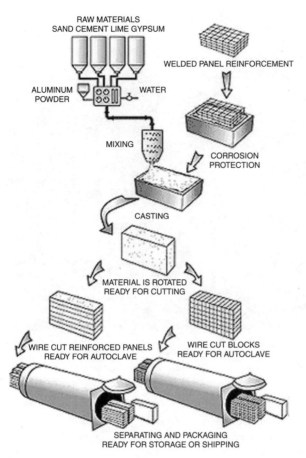

RAW MATERIALS
SAND CEMENT LIME GYPSUM

WELDED PANEL REINFORCEMENT

ALUMINUM POWDER   WATER

MIXING

CORROSION PROTECTION

CASTING

2-6 *Schematic of the operations of a plant for manufacturing autoclaved, aerated concrete.* Aerated Concrete Corp. of America

MATERIAL IS ROTATED READY FOR CUTTING

WIRE CUT REINFORCED PANELS READY FOR AUTOCLAVE

WIRE CUT BLOCKS READY FOR AUTOCLAVE

SEPARATING AND PACKAGING READY FOR STORAGE OR SHIPPING

Exterior walls

The exterior walls of buildings have been constructed with some form of concrete for hundreds of years. Concrete use in the walls of small buildings has risen sharply in the last fifteen years because of technology improvements and rising demand for the benefits concrete walls offer.

Concrete masonry is one of the most tried and true methods of constructing concrete walls employed today. These larger units are sometimes referred to as *concrete block.* Concrete masonry is a dry-cast masonry product produced economically in large quantities and stacked with mortar at the job site. It has been used for the walls of buildings for a hundred years. There are trained masons capable of installing it in almost any area. It has widespread acceptance. It can now also be made with a wide range of colors and premium finishes, for an attractive wall surface without extra installation steps (Figure 2-7).

2-7 *Building construction with concrete masonry walls.* Portland Cement Association

Precast wall panels several inches thick have been used for decades. New panel designs directed at smaller buildings over the last 20 years have made them even more economic in that application. Some of these newer panels include quality insulation, electrical chases, or studding to ease the job of finishing. A wide range of attractive finishes can be cast onto the surface (Figure 2-8).

Cast-in-place walls with removable forms have come a long way from their origins. Used mostly for foundations in the past, they are now growing rapidly in above-grade applications. This stems partly from advances in the formwork and accessories that provide such features as insulation and studding. The range of finishes that may be applied to the panels has grown as well (Figure 2-9).

Cast-in-place walls with stay-in-place forms have taken off with the development of *insulating concrete forms.* They are relatively flexible and have good insulation and built-in studding (Figure 2-10).

Tilt-up wall panels have been used in large, rectangular buildings with a simple finish for fifty years. But in that time there has been a steady advance in the construction process and the development of quality insulation and finish systems. Today many tilt contractors can economically produce buildings that are smaller and have complex designs and premium finishes (Figure 2-11).

2-8 *Installing precast wall panels.* Dukane Precast, Inc.

2-9 *Concrete walls constructed with removable forms.*

2-10 *Walls under construction with insulating concrete forms.* ICF Builders

2-11 *Tilt-up wall panels lifted into position.* Tilt-Up Concrete Association and Steinbicker & Associates, Inc.

2-12 *Walls under construction with autoclaved, aerated concrete.*

Autoclaved aerated concrete has been widely used in Europe for decades and is now coming into use for small buildings in North America. The material is lightweight because it is filled with many tiny air bubbles. It is available for wall construction in the form of blocks (for use as masonry) or panels (handled like precast panels). It offers good flexibility and integral insulation (Figure 2-12).

Floors and roofs

More and more concrete floors and roofs are finding their way into smaller buildings. They were once considered higher strength and quality than necessary and too expensive. However, several advances have produced floor and roof systems that use materials more efficiently than in the past, reducing costs. Buyers have also gained interest in the concrete decks because of their potential for greater strength, disaster resistance, and energy efficiency.

Cast-in-place floors and roofs with removable forms are a good complement to removable form walls. The floor forms are designed to connect to the wall

2-13 *Placing concrete for a floor built with removable forms.* Wall-Ties & Forms, Inc.

forms. The forms for both walls and floors are set together and the concrete is placed for both in one operation. The result is a quality, high-strength concrete floor at a relatively small incremental cost and time (Figure 2-13).

Insulating concrete form floor and roof forms arrived about five years ago in North America from Europe. Made principally of foam and light-gauge steel, the forms are readily cut into desired shape before the concrete is cast on top. They offer high insulation and easy connection points for attaching finish materials (Figure 2-14).

Composite steel joist systems employ high-strength steel joists that hold up temporary formwork during placement of the concrete, without shoring. The concrete bonds with the tops of the joists to make a high-strength floor. The open joists below permit flexible routing of utility lines (Figure 2-15).

Precast hollowcore planks are prefabricated at the plant and shipped to the job site. There they are set into place to create a quality floor or roof deck in only a few hours (Figure 2-16).

Autoclaved, aerated concrete planks are also shipped to the job site and set into position in a short time. The AAC material reduces the weight of the floor or roof and adds insulating properties (Figure 2-17).

2-14 *Assembly of the formwork for a floor built with insulating concrete forms.*
Seadore Masonry Inc.

2-15 *Composite steel joists in position.* Southeast Florida Polysteel

2-16 *Precast hollowcore planks hoisted in position for a floor.* The Spancrete Group, Inc.

2-17 *Topping walls with AAC floor planks.*

Exterior finish

The wear and durability of concrete makes it a natural for exterior finishes. More than many other finish products, ones made of concrete stand up to the elements and require little maintenance.

Stucco has been a widely used troweled finish for exterior walls for hundreds of years. Made with modern cement, it is highly durable. It is also flexible enough to cover almost any irregular wall feature or trim shape. It can be colored or painted and has a classic, time-honored appearance (Figure 2-18).

Concrete brick is a newer option. It offers a wall finish similar to traditional clay brick, but with great product consistency, a wide range of colors and finishes to choose from, and low waste (Figure 2-19).

Fiber-cement siding is pressed into shape in the factory to reproduce the look of any lapped wood siding products accurately. Yet it has significantly lower cost, greater consistency, greater durability, and lower maintenance than wood (Figure 2-20).

Manufactured stone provides a true stone masonry appearance at a much lower cost. It consists of highly accurate reproductions of natural stone. Manufactured stone is made with concrete that is cast into special rubber molds.

2-18 *Applying stucco to a wall.* Portland Cement Association

2-19 *House with a concrete brick veneer.* Portland Cement Association

The pieces are individually stained with natural tones. They are designed with flat backs to be adhered to the wall instead of stacked (Figure 2-21).

   Concrete roof tiles provide a premium, durable roof covering at a reasonable cost and with a wide range of appearances. They can closely match the shape and color of clay roof tiles, which are popular in many parts of North America. However, with proper molding and coloring they can also have a range of other looks, such as slate, other natural stones, and many unique, high-tech looks (Figure 2-22).

2-20 *Installation of fiber-cement siding.* CertainTeed Corporation

2-21 *Applying manufactured stone to a wall.* Eldorado Stone

2-22 *Installing concrete roof tiles.* Monier Lifetile, LLC

Landscape products

The durability of concrete also makes it a natural on the lot. As products for
landscaping are developed, concrete is replacing other materials. This process
is accelerating as the new aesthetic treatments make concrete landscape products highly attractive. Many buyers are even using these products to create
walls and walkways that are not strictly necessary. Like a garden, these products are there because they make the lot more beautiful. Using concrete in
landscaping is becoming so significant a part of American homes that the industry has coined a new term for it, *hardscaping.*

Concrete flatwork consists of slabs about 4 inches thick cast on the ground. It
has been used to create durable walks and drives for decades. Now the many
new concrete coloring and finishing techniques give it new options for beauty as
well. It is also available in a *pervious* version, which allows water to drain through
into the soil, reducing unwanted rainwater runoff (Figure 2-23).

Interlocking concrete pavers are also used for paving, but with individual
masonry pieces molded in a dry-cast process. They are set much like traditional paving stones, but they are more precise, less expensive, and have a
much wider array of shapes, colors, and possible patterns. They are now also
available in pervious versions that allow rainwater to drain into the soil below
(Figure 2-24).

2-23 *Installation of concrete flatwork.*

2-24 *Workers installing concrete pavers.*

Segmental retaining walls are large concrete masonry units precisely dimensioned and outfitted to create walls holding back the earth economically. Compared with the older treated wood timbers, they are more durable, lower in maintenance, and capable of being produced with the look of natural stone

2-25 *Stacking up a segmental retaining wall.* Landscape Development, Inc.

or any of a range of more novel, high-tech materials. They do this without the cost or inconsistency of natural stone (Figure 2-25).

Decorative concrete

Decorative concrete is not really a product, but a set of techniques for giving the surfaces of any concrete product color, textures, and patterns. Its use has exploded over the last twenty years. It is being applied particularly to give striking appearances to countertops, concrete floors, and exterior flatwork.

Over the last two decades, many new concrete pigments, chemical treatments, and finishing methods have appeared. Craftspeople all across North America have discovered and taken advantage of these to create surfaces with distinctive and lasting beauty. Interest and sales are growing rapidly.

One of the key attractions of architectural concrete elements is that almost every one is unique. They are custom products. The craftsperson creates a distinctive look for each customer or building. A wide range of appearances is possible. In some ways they are actually artwork or fashion. As one might expect, they are highly prized, high-margin amenities.

2-26 *Application of decorative features to concrete flatwork.* Portland Cement Association

There is a broad range of decorative techniques, applied in various ways. Decorative concrete countertops are created after the cabinetry that will hold them is in place. This helps achieve a precise fit. They may be constructed in the controlled environment of a shop or cast in place on top of the cabinets. The decorative surface of a floor can be created on-site with any concrete floor. This is usually done after the concrete has cured and most of the rest of construction is complete. Some types of decorative features are applied to exterior flatwork immediately after casting, while the concrete is still wet. Others come after the concrete has cured (Figure 2-26).

The specific decorative concrete techniques now available are described in detail in Chapter 32.

3

Materials

The many valuable products and systems described in this book are made possible by the chemistry of concrete and the technology of related materials. Understanding basics of these materials is helpful to understanding the products. Readers already familiar with these fundamentals may wish to skip this chapter or parts of it.

Origins of cement and concrete

Modern concrete is made possible by *portland cement*. Cement and concrete are different and are often confused.

The word "cement" is broadly used to refer to almost any adhesive that binds pieces of solid materials to one another. In construction, the term cement is shorthand for the particular cement called *portland cement*, a gray powder made by pulverizing and heat-treating a precise mixture of minerals. It hardens into a rock-like mass when mixed with water. Technically, an engineer would say that it *gains strength*. Once a mass of cement has hardened, it is virtually unaffected by common heat or cold, water or dryness, sunlight, most chemicals, or fire.

The Romans and some older civilizations made an earlier type of cement that consisted mostly of limestone and clay reduced to powdered form. The Romans got some of the necessary minerals by crushing certain volcanic stones.

The modern version of cement came from John Aspdin, a mason in England, in 1824. Through experimentation he developed a recipe and a process that yielded a cement with greater strength and consistency than any that came before. He named it portland cement because it looked something like the limestone from nearby Portland, England. It is now made all over the world and sold under many different brand names by many different companies. The name is not capitalized because portland cement is now a generic term, like "high-carbon steel" or "treated lumber."

The general term *concrete* simply means a mass consisting of solid pieces bound together with some adhesive or cement. In construction it is used to

3-1 *The major ingredients of construction concrete: sand, stones, cement, and water.*
Portland Cement Association

mean a mass of pieces bound together with portland cement. The pieces are referred to as *aggregates*. They are almost always sand and stones, but technically portland cement mixed with any of a wide range of other materials is still a form of concrete (Figure 3-1).

When the materials of concrete are mixed with water, the cement binds the whole mass together. The concrete becomes hard or gains strength much like cement alone does, and exhibits similar properties of strength and durability. However, concrete is considerably less expensive because sand and stone are available at a lower cost than pure cement. Using aggregates also gives the user some flexibility. It is possible to vary some of the properties of concrete by mixing in different types and proportions of aggregates.

Strength and reinforcement

Concrete is relatively high in *compressive strength*. In simple terms, it can bear great weight or loads without shattering (Figure 3-2).

Plain concrete is lower in *tensile strength*. In other words, it cannot bear as great an amount of pulling from each end without breaking apart.

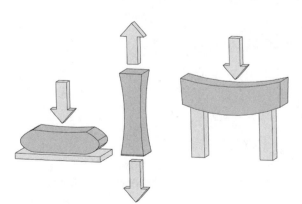

3-2 *The basic forces of compression (left), tension (center), and bending (right).*

This is important because construction elements must endure both *compression* (pressure) and *tension* (pulling). A wall must hold up the building and everything on it without collapsing. Yet it must also hold the roof down without breaking apart when wind gets under the eaves and lifts up. Likewise, a floor or roof must endure the pushing and pulling that results from when wind pushes or pulls on the outside walls.

Construction elements must also have *bending strength*. They must not split when forces try to bend them one way or another. However, bending is simply a combination of compression and tension. Bending an object simply compresses one face and pulls on the other. If the material has both good compressive strength and good tensile strength, it will resist the bending without breaking. In a building, the wind and ground pressure on a wall tend to bend the wall inward or outward. The weight on a floor or roof tends to bend it down.

In all these situations, if concrete walls and floors and roofs are to hold up they may need added tensile strength. The limited tensile strength of concrete alone may not be enough. The extra tensile strength is generally achieved by adding *reinforcement* to the concrete. Reinforcement is some form of bar or fiber embedded in the concrete that has high tensile strength. When forces try to pull the concrete apart, the resistance of the reinforcement holds it together.

As an example, consider a stack of books. They create a sort of column that can support the weight of a bowling ball. The column has some compressive strength. If you pull up on the top book, they separate easily. The column has no tensile strength. Now wrap several rubber bands around the stack. If you try to pull the books apart again, the bands resist and the column stays together. The bands act as a form of reinforcement.

When designing a concrete building, engineers select the reinforcing that is needed to give the walls all the tensile and bending strength they need. Put another way, they assume the tensile strength of the concrete is zero, and make sure there is enough reinforcement to provide all the tensile strength that the wall needs. They count on the concrete only for its compressive strength. This method of engineering is called *reinforced concrete design*, and it is used today for most structural concrete elements of a building.

Cement

Portland cement is available in different varieties. Changing the proportions of the minerals that go into the cement and adjusting the production methods can shift its properties (Figure 3-3).

Types

An infinite number of different cements are possible, but there are now established categories of cement, called *types,* that are defined by industry stan-

3-3 *Portland Cement.* Portland Cement Association

dards. The cements included in any one of these categories must adhere to certain minimum properties. These types include most of the cements used in construction today.

Type I cement is the most common and usually the least expensive type used in construction concrete. It normally shrinks very slightly during hardening.

Type II or "low shrinkage" cement shrinks substantially less than Type I. It is typically more expensive and gains strength more slowly.

Type III or "high early" cement also has reduced shrinkage and hardens quickly.

White cement is not one of the officially defined Types, but it is important. It is produced without iron so that it is nearly white instead of gray. White cement is necessary for concrete that needs to be of very light or bright color. It is generally more expensive than the gray cements.

Strength gain

The hardening process of cement is also called *hydration* or *curing*. It occurs because the minerals of cement, when they are in the presence of water, form crystals that link to one another. This creates a single hard crystal structure, something like bits of carbon bonding into one diamond when they are under great heat and pressure.

The longer the cement remains moist, the more links are formed and the harder the mass becomes. However, the strength gain gradually slows. After a few weeks the mass is nearly as hard as it will ever get. Any additional gain in strength occurs very slowly.

The water is not a part of the crystals. In fact, it evaporates off later and the concrete does not lose any strength as a result. Water acts is a catalyst, a substance that causes a chemical reaction to occur by helping other materials combine, but it does not combine with them.

The process is sensitive to the amount of water present. If there is too little water, not all of the possible crystals form, and the final mass remains weak and crumbly. If there is too much water, the cement particles can separate from one another as they float in water. The result is small, individual crystals that will blow apart like a powder.

Engineers have determined that a reliable measure of whether a batch of cement or concrete has the proper amount of water for achieving its potential strength is the *water-cement ratio*. This is sometimes written *w/c*. The water-cement ratio is the weight of the water in a batch of concrete divided by the weight of the cement itself. For example, a batch containing one pound of water and three pounds of cement would have a w/c of 0.33.

Cement or concrete can cure fully when the w/c is 0.28 or more. However, with these proportions the mix may be impractical to use. It will be very stiff and difficult to get to fill a form or mold completely. If any water evaporates, the amount left in the mix will drop below the minimum and full strength may not be achieved. Using such "dry" concrete is generally practical only in certain manufacturing processes, where special equipment handles the stiff mix and environmental conditions can be controlled to keep the material moist.

In the field and in other manufacturing processes, the mix is usually prepared with a w/c of about 0.4–0.5. This is usually enough water to make a mix that is fluid enough to work with. It is also enough to cause full crystal formation, even when there is modest evaporation. As the w/c is raised above 0.5, the strength of the resulting material begins to fall significantly.

The hydration reaction of cement and water produces heat. Scientists refer to it as an *exothermic* reaction. This may be a consideration in handling and placing concrete correctly.

Portland cement substitutes

Some other powders react much like portland cement when mixed with water. They form hard masses with great strength and resistance to environmental conditions. As a group, these are often referred to as *cementitious materials*.

Included are the volcanic minerals the Romans used for their version of construction cement. Also included are other modern industrial byproducts like blast furnace slag and *fly ash*, which is left over from burning coal. In fact, these industrial byproducts are in some ways quite similar to the Roman volcanic cement materials. All of them are grouped together under the name *pozzolanic cements* or *pozzolans*. The term "pozzolan" comes from the name the Romans used for their volcanic material.

Today pozzolans, especially those from industrial processes, are frequently blended with portland cement for use in concrete. When added in limited quantities they can give the mix some desirable properties. As waste products, they are usually also inexpensive so they can reduce the total cost of the mix. Their use also takes an unwanted byproduct out of the waste stream.

Concrete

The concrete most of us are familiar with is a mixture of portland cement, sand, and small stones mixed together with water to harden. The contractor ordering a batch from the local ready-mix plant typically must specify three different properties for the mix. These are the compressive strength, aggregate size, and slump.

Compressive strength is the maximum pressure a material can withstand before it shatters or collapses. Concrete compressive strength can be varied over a wide range by adjusting ingredients. Most concrete used in construction has a compressive strength of 2,000 pounds per square inch (psi), 2,500 psi, 3,000 psi, or 4,000 psi. In Canada, where the metric measurement system is used, the common concrete strengths are 15 MPa, 17.5 MPa, 20 MPa, and 30 MPa.

Using concrete of correct strength is critical in most construction applications. For structural parts of a building, the code or an engineer determines the strength necessary. The contractor does not have authority to deviate from that. This applies to the concrete the contractor orders from the ready-mix plant and places in the field. It also applies to the concrete strength of the manufactured products that the contractor orders from the factory, such as concrete masonry or precast panels.

Concrete strength is also important for items that are not part of the building or are nonstructural, such as concrete brick, pavers, flatwork, and so on. This is because the stronger the concrete, the more durable and resistant to damage it generally is.

Aggregate size is the dimension of the largest pieces of aggregate in the concrete mix. Most common construction concrete has an aggregate size of ¾-inch (19 mm in metric). That means that the largest stones in the mix are to be no more than ¾ of an inch across at their longest point. A maximum aggregate size may be specified by the code or the engineer to help insure that the concrete can flow completely around the rebar or into all spaces of the forms. However, it is instead often left to the contractor (in the field) or the manufacturer (in the factory). The tradeoff they face is that concrete with smaller aggregate tends to flow more easily, but it also tends to be more expensive.

Slump is measured by filling a foot-high cone with concrete, turning it over, and measuring how far the concrete "slumps" down from the original 12-inch height. Engineers recommend taking the slump of different batches of concrete to determine whether they are consistent. If two batches exhibit sharply different slumps, it is clear that they were not mixed with the same ingredients in the same proportions. However, contractors and manufacturers generally use slump as a rough indicator of how easy the concrete will be to work with (Figure 3-4).

Higher slump concrete tends to flow into forms or molds more easily and with less prodding. It is also easier to trowel it smooth, and sometimes that is important. Some refer to the ease of working with concrete as *flowability* or

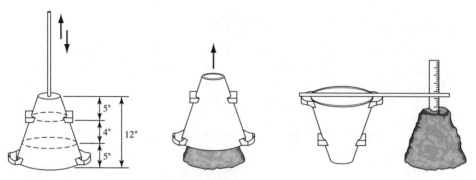

3-4 *Testing the slump of concrete with a slump cone.*

workability. Slump is not a totally reliable indicator of this because other factors such as aggregate shape and size influence it as well. However, for many it is a handy indicator.

Since slump itself is not a determinant of the strength or durability of concrete, engineers and codes do not generally specify slump. The contractor or concrete product manufacturer may select a slump, so long as they adhere to the critical requirements such as the strength specification.

Strength

Wet concrete gains strength over time. It is critical that it remain moist to achieve the required strength (Figure 3-5).

Since the strength of concrete is a moving target, it is necessary to be specific about the point in time used to measure it. By general agreement, the

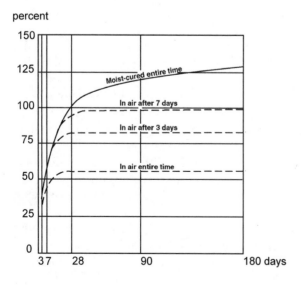

3-5 *The rate of strength gain of concrete.*

strength that a batch of concrete reaches after remaining moist for 28 days is considered its *full strength*. It is also sometimes called the *design strength* of the concrete. When a ready-mix plant receives an order for concrete with strength of 3,000 psi, it prepares a mix that it knows will reach strength of at least 3000 psi in 28 days so long as it remains moist. The concrete can reach even greater strength later. In fact, it may continue to gain strength slowly for many more months if some water remains present. However, it will still be classified as "3,000 psi concrete."

The concrete may fail to reach its full strength if it is not properly handled. If it is allowed to dry out or freeze, the mix is robbed of the necessary moisture. If that happens too early, the concrete will stop hardening and peak out at some lower strength. Similarly, if extra water is added to the mix, the water-cement ratio rises and the concrete may never reach full strength either. These can be serious problems.

For wet concrete placed in the field, the contractor is responsible for ordering the correct concrete strength and handling the material so that it achieves its full strength in the field.

For concrete products produced in a factory, the manufacturer generally guarantees a particular final strength of the product. The manufacturer may use different concrete mixes, cure times, and cure methods so long as they all achieve this advertised strength.

Consolidation

If concrete is simply dropped into a form or mold and allowed to cure, it will often contain large trapped air pockets called *voids*. Voids deteriorate the properties of the final concrete. The concrete may be significantly weaker, its grip on any reinforcement is reduced, and its surface may be uneven and unattractive.

To reduce voids, the installers *consolidate* concrete. Consolidation is temporarily vibrating the concrete to make it more fluid. This enables the air to bubble to the top and out of the mix. The favored method of consolidating concrete cast at the job site is with an electrically powered device called a *vibrator*. It consists of a long, flexible shaft that is usually 1 inch to 2 inches in diameter and a motor that causes the shaft to vibrate rapidly. Workers insert the shaft into the concrete, and the air bubbles up, almost as it does from boiling soup. In factory-produced products, the consolidation is often done instead by vibrating the mold or forms around the concrete (Figure 3-6).

During consolidation, the pressure the concrete puts on the forms or the mold around it increases sharply. If the vibrations are too strong or they are maintained for too long, they may damage the forms or mold or push them out of correct position. It is important to have knowledgeable operators handling the equipment.

3-6 *Using a vibrator to consolidate concrete.* Meyer Brothers Building Co.

Segregation

Segregation is the uneven distribution or "clumping" of aggregates in concrete. Where it occurs it can be a problem. A batch of concrete is mixed with select ingredients in select quantities to achieve the desired strength and durability properties. If the stone and sand end up concentrated in one section of the concrete, there will be areas of the final material where it is weaker and lower in durability than intended.

The aggregates of concrete are relatively heavy, so gravity tends to pull them down to the bottom of the mix. However, conventional construction concrete is thick enough that aggregates will not fall much while the concrete is sitting still. Segregation more often occurs when concrete is consolidated too long in one place. The vibrations make the concrete fluid enough that the aggregates can sink.

For concrete products produced in a factory, manufacturing procedures are set to avoid segregation. For concrete placed in the field, the contractor is responsible for using installation and consolidation methods that avoid segregation.

Floating

Flat surfaces of concrete that are installed wet in the field are often *floated*. Floating is repeated pressing or stroking of the surface with a flat tool to make the surface smooth and compact the material. This can also help to harden the surface and make it more durable.

Floating is most often done with concrete flatwork (floors, walks, and drives) and stucco. It can be done with a hand trowel. However, for large expanses there are larger tools called *floats* that make work faster and, in the case of flatwork, allow the installer to stand. Power tools are also available, but they are rarely used on small projects.

Cracking

Most concrete shrinks slightly as it cures. For a large section such as a wall or floor or driveway, this may lead to cracks forming.

Frequently cracks are not important. For example, the concrete in most walls is reinforced by steel reinforcing bars. If small cracks form in the concrete, the rebar holds the wall together

In other situations it is important to *control* cracking. *Crack control* amounts to taking measures that will keep any cracks small or limit them to certain areas where they will not be a problem.

One traditional method of crack control is intentionally creating joints or lines of weakness in the concrete. This leaves straight, predictable joints that can be treated so they are not a problem.

A wall might intentionally have a narrow gap about every forty feet along it. This is called a *control joint*. It breaks the wall up into smaller sections that can each shrink slightly without creating random cracks. The concrete may shrink away from the control joint slightly and increase the size of the gap. However, this need not be a problem. A variety of elastic products are available to seal this joint so that air and water do not penetrate.

A concrete sidewalk usually has a line indented in it about every three feet. If the sidewalk cracks it usually does so along this line of weakness. The sidewalk becomes a set of large concrete squares. Small fragments do not form, and the walker can easily see the straight, regular joints.

A method used increasingly to control cracking is reinforcement. Reinforcement embedded in the concrete limits how far it can shrink. This in turn limits the length and the width of any cracks that form.

Admixtures

Admixtures are chemicals mixed into concrete in small amounts to give it special properties. There are many admixtures for many different purposes. A few are common enough that background on them here is useful.

Plasticizers make concrete more fluid so that it flows more readily. This is extremely useful when the concrete is to be cast over large areas or into tight spaces. With plasticizers added, the concrete flows more readily and less work is required to move it to its desired location and get it to fill completely. This method of making concrete fluid is usually far superior to the shortsighted trick of simply adding water. By adding water, the contractor or producer is decreasing the strength and durability of the final concrete. Unless careful track is kept, the concrete may be too weak to meet specifications or perform its function. But plasticizers generally add cost, so selecting when, how much, and what type to use is a sensitive decision.

Accelerators increase the speed of concrete's strength gain. They are often used in concrete that will be exposed to cold weather to get it to harden sufficiently before it freezes. However, accelerators are generally not recommended for concrete that will contain steel reinforcing. Common accelerators can corrode steel and certain other materials.

Air entrainers trap tiny air bubbles in the concrete. Small quantities of these bubbles make concrete more resistant to freezing and thawing. This has been a consideration for outdoor flatwork, such as driveways and sidewalks. In cold climates flatwork experiences many cycles of standing water freezing on top. More recently, advanced air entrainers have been used to make concrete with over 50 percent air bubbles. These provide light-weight, insulating ability, and other properties that are desirable in some uses.

Pigments give concrete a desired color. The pigments used in concrete are mostly metal oxides. These have the advantage that they are durable. Through extreme weather and aging, they remain embedded in the concrete and retain their color. Almost any color is possible by mixing together the correct combination of pigments. To achieve a light or bright color, it may be necessary to use white cement as well.

Varieties

Although most concrete contains sand and stones, technically many mixtures with other ingredients added to water and cement are considered concrete.

Grout, mortar, and *stucco* are concretes without stones. The only aggregate in the mix is sand. Grout is used to fill small cavities like the cells of concrete masonry. Leaving out the stone allows it to flow in more rapidly and fully. Mortar is placed in a thin layer between masonry units such as bricks, blocks, and stones to level them precisely and adhere them to one another. Larger stones could interfere with accurate leveling. Stucco is troweled smooth over a surface to finish it. Stones might make it difficult to create a thin, precise layer of material (Figures 3-7 and 3-8).

Fiber-cement is a mixture of wood fibers and cement. It can be shaped and is durable much like concrete. However, the fibers make it more pliable.

3-7 *Pumping grout into a wall.* ICF Builders

3-8 *Placing masonry units in mortar.* Portland Cement Association

Thin sections of fiber-cement can flex some without breaking. A nail or screw can be driven through without splitting the material.

Autoclaved, aerated concrete (AAC) uses air entrainers that make the final product mostly air by volume. This lightweight concrete resembles a fine pumice. It is much lighter than conventional concrete and it has a much higher insulating value. To produce it, wet concrete without stones is heavily aerated and then cured in an autoclave. This is a steam chamber. It keeps the concrete moist, warm, and under pressure to speed the curing process and control the aeration of the final material.

Self-compacting concrete (SCC) is a new variety that may have important benefits. It is also called *self-consolidating concrete*. It contains admixtures that make it flow readily to fill up formwork completely without significant voids. The work of manually moving concrete into position and consolidating it is sharply reduced (Figure 3-9).

The formulas for SCC call for carefully sized aggregates, highly effective plasticizers called *superplasticizers*, and an admixture called a *viscosity modifying agent* (VMA). The superplasticizers increase the slump of the concrete so much that it is difficult to measure. The concrete spreads out like honey. The highest point of the "pile" is simply the top of the largest stone. Instead of measuring slump, users get an indication of the fluidity of the concrete by measuring the *spread*. This is the diameter of the circle of concrete that forms

3-9 *Self-compacting concrete spreading to create a floor.* Lafarge North America

after one minute. The spread of self-compacting concrete is typically 25 inches to 30 inches. The viscosity modifying agent prevents the concrete from flowing too rapidly. Its flow is more like honey than water. It also insures that the aggregate moves evenly throughout the mix as it flows, instead of falling to the bottom. This prevents segregation. Self-compacting concrete is more expensive than conventional construction concrete. As it is a new product, building professionals are gradually gaining familiarity with SCC and determining where it can be best used.

Reinforcement

Concrete reinforcement is available in several different forms.

Rebar

Steel reinforcing bar, also called *rebar*, consists of steel bars. These are set inside the forms or molds. The concrete is placed around the rebar and encases it (Figure 3-10).

Most rebar is *deformed.* The manufacturer creates ribs on the surface of the bars so that the concrete grips it securely.

3-10 *Steel reinforcing bar.* Tilt-Up Concrete Association

Rebar is available in different *diameters* and *grades* to provide different strengths of reinforcement. In the United States the bar is available in diameters with increments of one-eighth of an inch: $\frac{3}{8}$ in., $\frac{4}{8}$ in., $\frac{5}{8}$ in., and so on. For convenience, rebar with a specific diameter is referred to as a particular "number" of bar. Rebar with a $\frac{3}{8}$ in. diameter is called "number 3" rebar; rebar with a diameter of $\frac{4}{8}$ in. is called "number 4" and so on. Larger diameter bars have greater total tensile strength than smaller diameter bars. Therefore it may be necessary to use fewer of them to achieve the same strength in the final concrete object. They are also more expensive and may be more difficult to work with because they are heavier and harder to bend. In small buildings almost all the bars are number 3, 4, or 5. Large buildings and heavy construction often use larger diameters.

The grade of rebar refers to the strength of the steel material it is made from. "Grade 40" rebar is made from steel with a tensile strength of 40,000 pounds per square inch (psi). In other words, a bar made of grade 40 steel that is one square inch in area (about $1\frac{1}{8}$ in. in diameter) could be pulled with a force of 40,000 pounds before it would snap. The steel in "grade 60" rebar has a strength of 60,000 psi. A smaller amount of higher-grade rebar is needed in a section of concrete to give it the same strength. However, it is also more expensive and may be more difficult to bend.

Rebar may be ordered to any length, or purchased in standard lengths. It may be ordered with bends formed in it. It may also be cut and bent at the job site with special tools.

The position of rebar in concrete is important, and the contractor is responsible for it. The rebar is positioned where the wall will be most subject to tensile (pulling) forces. However, it must not be close to the surface of the concrete or else there will not be enough concrete around it to grip it properly. The building code or construction documents specify where particular pieces of rebar need to be located.

The rebar must also be secured so that it does not shift when concrete is placed around it. This is done by various means. For a vertical structure, such as a cast-in-place wall, there are usually crosspieces running through the cavity in the formwork. It is convenient to secure the rebar to these with wire (called *tie wire*) or plastic zip ties. For horizontal structures, such as cast-in-place floors or walls cast in the horizontal position, the rebar must be elevated an inch or more above the bottom of the form. This is usually accomplished by setting the bars on devices called *rebar chairs* or just *chairs*. They are small supports set on the bottom of the form every few feet under the locations of the bars. Chairs are available in many different heights to position the rebar at any desired depth in the concrete.

Prestressed steel reinforcing strands

Steel reinforcing strands are similar to a steel cable. They consist of many steel wires wound together much like fibers wound to create rope. They are frequently used as reinforcing in precast products with a procedure known as *prestressing*.

Under prestressing, the strands are run through the forms and are stretched by pulling on either end. The concrete is cast around them and embeds them. After the concrete cures, the strands are released and their ends cut off.

Prestressing is quite efficient reinforcing. Since the strands are already partially stretched, they hold the concrete together tightly. Consider the example of a stack of books held together with rubber bands. If the bands are slightly loose, it will still be easy to pull the books a significant distance apart. If they are very tightly wrapped, it will be difficult to pull them apart.

Welded wire mesh

Welded wire mesh consists of small-diameter steel bars, called "wires," arranged in a grid and welded together where they cross. Manufacturers offer it in flat sheets or in rolls. It is used mostly for reinforcing flatwork such as floors and outdoor flatwork to limit cracking. Because the bars run in two directions and are usually spaced close together, any cracks that form will be close to reinforcement. This prevents them from becoming large. It is also possible to reinforce concrete for crack control by installing many individual bars instead of mesh. However, installing mesh takes less time and labor (Figure 3-11).

3-11 *Welded wire mesh.* Tilt-Up Concrete Association

Mesh is available in many different varieties. There are several possible wire diameters. The spacing of the wires in the mesh also varies. In general, mesh with greater diameters and closer spacing provides more strength. Mesh with smaller diameters or wider spacing is usually less expensive and easier to work with.

Mesh is almost always installed in concrete structures that are cast horizontally (slabs on grade, floors, etc.). They are set on chairs to raise them to the desired height.

Fibers

Reinforcing fibers are thread-like strands with high tensile strength mixed into concrete. Traditionally they are used as a replacement for welded wire mesh to control cracking. However, some fibers used in relatively high amounts are being used to provide structural strength to concrete and reduce the amount of rebar required.

The fibers most used today are made of steel or polypropylene plastic. These are sometimes called *synthetic fibers* because they are man-made materials. This contrasts, for example, with the wood fibers used to make fiber-cement products. They can be mixed into concrete in different concentrations, called *doses.*

Using fibers reduces or eliminates the time and labor to install mesh or rebar. If the fibers are installed at the correct dosage and mixed in thoroughly they are located at very close intervals, without significant gaps and in all directions. This is helpful in closely limiting cracking. These features are weighed against the cost of the fibers, their unfamiliarity to some in the building industry, and any effects they might have on how readily the concrete flows and can be handled.

Foam

Rigid plastic foams have become so widely used in concrete construction systems that users need to be familiar with the major types and properties. They are used mostly as insulation in walls, floors, and roofs. They are also sometimes the least expensive and quickest way to make a form for unusual concrete features. This is because they are easily cut and glued together to create even complex and intricate shapes (Table 3-1).

Expanded polystyrene

Expanded polystyrene (EPS) is created from "beads" of a plastic called polystyrene infused with many tiny gas bubbles. When the beads are heated in a mold, the plastic softens and the gas expands. The beads expand to fill the mold completely and fuse to one another. The result is lightweight plastic foam in almost any shape desired. It is commonly used to make disposable coffee cups and the protective guards placed on the corners of a new television set in the box.

Table 3-1. Approximate Values for Properties of Plastic Foams Used in Construction.*

	R-value per Inch	Compressive Strength**	Tensile Strength	Flexural Strength	Flame Spread	Smoke Developed
EPS	4	25 psi	22 psi	58 psi	10	125
XPS	5	33 psi	60 psi	55 psi	5	165
Poly-urethane	6	30 psi	30 psi	—	20	250

*All values may vary significantly with such factors as the density of the foam, the manufacturing methods, and any additives.
**To 10 percent deformation.

Expanded polystyrene is one of the least expensive plastic foams. It has a good R-value, and is easy to cut and glue.

Expanded polystyrene can be produced with different *densities*. The density of EPS is its weight per cubic foot. Most EPS used in construction has a density of 1.0 pounds per cubic foot (pcf), 1.5 pcf, or 2.0 pcf.

Denser EPS contains more plastic and less air. The same volume of a denser EPS has a somewhat greater insulating ability (a higher R-value) than a less dense EPS. It is also stronger. However, it is more expensive because it contains more plastic. Higher-density EPS is used where the greater strength or R-value is important enough to justify the higher cost.

Extruded Polystyrene

Extruded polystyrene (XPS) is also produced from polystyrene plastic and expanding gas, but by different methods. The ingredients are mixed in a heated chamber to form a bubble-filled liquid that is extruded into sheets of foam. These sheets are commonly given a color that is the trademark of its manufacturer: blue, pink, or green.

Extruded polystyrene is available in sheets of a variety of thicknesses. Common thicknesses used in construction range from about half an inch to three inches. It is not as readily produced in complex shapes as some other foams are. However, XPS is also easily cut and glued to create new shapes.

Nearly all XPS sold today has a density of 2.0 pcf. It has a somewhat higher R-value and greater strength than common EPS. It is also generally more expensive. How great these differences are depends on exactly the densities and varieties of the XPS and EPS being compared.

Polyurethanes

Polyurethanes are actually a category of related foams. They do not all have exactly the same chemical ingredients, but they are similar. Polyurethane foams are formed by a reaction of two different chemicals. The reaction produces a plastic as well as a gas that froths up the mix. After the plastic cures, the result is rigid plastic foam.

Polyurethanes are available in sheet or molded form. One of the polyurethanes commonly used in construction is an insulation board faced on one side with foil.

Polyurethanes have a higher R-value than any of the other plastic foams. They are also usually the most expensive, so their use is concentrated in applications where it is important to get a high R-value in a limited thickness. They can also be manufactured in a variety of densities, with somewhat different properties in the material of different densities.

Fire properties

Existing codes and standards require foams used in construction to have certain properties related to fire resistance. Virtually every manufacturer of foam or a foam product used in construction includes ingredients to meet or exceed these requirements.

The fire resistance standards include two major requirements. The first is for *flame spread.* This is measured by applying a controlled flame to the material and measuring how far the flame spreads. The scale for flame spread is set by red oak. The distance that the flame spreads when it is applied to red oak is considered to be 100. A material on which the flame travels twice as far is said to have a flame spread of 200. If the flame travels half as far, the flame spread is 50. Foam used in construction is required to have a flame spread of not over 75.

The other requirement is for *smoke developed.* This is a measure of how much smoke enters the air when the material is subjected to a flame. Its scale is also set by red oak, whose smoke developed is considered to be 100. Foam must have a smoke developed of not more than 450.

4

Background for evaluating concrete products

Parts 2–6 of this book describe important concrete products and systems in detail. Each of these products is different. However, the information for each product or system is divided into a set of standard topics. For most of these topics, knowing certain background information is helpful.

Market

Most products appeal more to certain types of buyers than to others. It is useful to divide buildings into different types, because buyers of the same type of building are likely to have similar preferences in the materials they buy. We commonly refer to these groups of buildings by type as *market segments.*

There is no one standard way to divide up buildings. A common breakdown is shown in Figure 4-1.

Any of these segments may be divided into finer segments, if that is useful. For example, among manufacturing buildings, wineries will have very different requirements from electronics fabrication plants. They may also be broken down by size. For example, small hotels will likely have different engineering requirements and be constructed with different materials than large ones.

It is also often useful to divide buildings according to their cost or the income or wealth of the buyer. Examples of such subsegments are upper-income homes or budget hotels. These things are important, because when the customers have more money, they are more likely to prefer products that offer greater benefits and to be willing to pay for them.

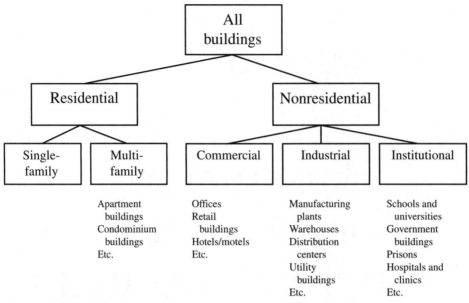

4-1 *A division of building construction into market segments.*

Cost

The cost is one of the most important and sought-after pieces of information for evaluating a construction product. It is also one of the most difficult to provide reliably. The costs associated with any construction product vary widely because the costs of the materials themselves and the labor to install them are different in different regions of the country, the building design and work conditions may be sharply different, and costs change significantly from time to time with inflation, material and labor availability, and general business conditions.

Another source of confusion with costs is exactly what cost or set of costs one is quoting. Figure 4-2 divides up the costs of a construction product or system.

Some price quotes include such miscellaneous items as fuel and equipment in the labor costs. In this case the materials and labor costs added together equal total installed cost. Some people therefore refer to total installed costs simply as "labor and materials."

A typical subcontractor's markup can be anywhere from 10 percent to 40 percent of the total installed cost. This compensates the subcontractor for such activities and costs as finding and ordering the proper materials, supervising the workers, coordinating with the client, phone, fax, taxes, and all the other expenses of running a business.

This book provides up to four different types of information about the cost of a product or system.

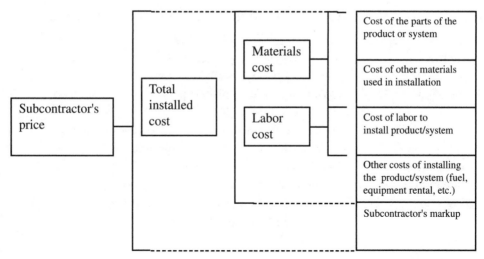

4-2 *Breakdown of the costs of a product or system in a building.*

First is how the cost of the product *compares to the costs of alternative products*. This is usually fairly consistent from time to time and place to place. For example, buyers almost everywhere agree that the total installed cost of fiber-cement siding is between the cost of vinyl siding and sawn wood siding. These rules of thumb are likely to hold true for most readers.

Second is the typical *range of subcontractor's prices* reported for installing the product or system. If reports of the total installed cost of a product are $2.00 to $3.00 per square foot, the reader at least knows what general range of cost to expect.

Third is a *table of the total installed costs* for a "typical" project and location using the system. This may not give the reader an accurate estimate of what the costs will be for a particular project. But it will give an approximation of how total cost breaks down into parts, and what a rough "midpoint" for cost of the product is.

Fourth is a description of *factors that heavily influence total installed cost*. These may include such factors as regional price or labor rate differences, aspects of a project that drive costs of using the product up or down, and so on. With these, a reader can judge whether costs of a specific project are likely to be higher or lower than the figures presented here.

Code and regulatory status

Table 4-1 lists the supporting materials commonly submitted to building departments to justify the use of a particular building product. For each of the products discussed in this book it is useful to know which materials are most commonly required, since this affects the amount of effort the designers and contractors

Table 4-1. Support Materials Submitted to Building Departments.

Support Materials	Notes
Custom engineering	Required for most large and some small buildings Almost always adequate
Code section	Available for established products and methods Usually adequate for small building
Evaluation report	Available for most established and some new products Often adequate for small buildings
Manufacturer's documents	Available for most established and new products Sometimes adequate for small buildings
None	Almost always adequate for parts of the building not regulated by the building code

will need to get their use of the product accepted. As a general rule, the newer the product and the newer the ways it is being used, the more time and money the builders may have to expend on getting building department approval.

The size of the project also makes a difference. More effort is required to satisfy the building department with the plans of larger, more complex buildings.

Consider how local building departments operate. Local building codes describe what can and cannot be built. Building designers are required to submit their plans for upcoming buildings to the department for inspection. During construction, the department sends employees called building inspectors to the job site to examine the work. If any of this reveals work that does not meet the rules, the building department can order work to stop until the violations are corrected.

The law usually gives the local building department and inspectors a great deal of discretion. It is often up to the inspector's judgment to decide whether a particular building practice meets the building code. The inspector may even decide that a practice is not acceptable even though it does not exactly violate the code. The judgment of the building official is especially important if a contractor uses a product or procedure that is not specifically discussed in the code.

Most local buildings codes today come from one source. Over a hundred years ago, when cities, towns, and counties began to pay more attention to how buildings in their areas were constructed, it became apparent that most of these jurisdictions did not have the time or money to write their own code from scratch. There arose nonprofit organizations designed to write a code for them. These organizations are called *model code agencies*, and their codes are called *model codes*. A city or town might pass a law that their building code would be, say, the Uniform Building Code, which was written by the International Conference of Building Officials. The city was said to *adopt* this model code. They might even adopt it with a few changes that they felt were necessary to make it fit their particular area better.

Today it appears that nearly all the localities of the U.S. will soon adopt the building codes of the *International Code Council* (ICC). Most have already done so or announced their intentions to do so. The ICC has written the *International Residential Code* (IRC), which covers homes, and the *International Building Code* (IBC), which covers larger and nonresidential buildings.

The situation is similar in Canada. Canada has long had a single model building code, the *National Building Code of Canada* (NBC). It is adopted by the Provinces and Territories, sometimes with changes. The NBC has different sections that cover small and large buildings.

Key objectives of the IRC, IBC, and NBC are the structural adequacy of buildings and the health and safety of their occupants and neighbors.

As listed in Table 4-1, the simplest situation for a construction product concerns a product or use that is not considered to be critical enough to be regulated by the building code. A good example is the aesthetic surface finish of most parts of a building. The texture of the floor surface, the color of the walls, and the like are generally not restricted because they are not considered to have a significant impact on such matters as safety, health, and welfare. In these cases, the building official will rarely ask for any information about the products involved.

The code becomes restrictive in matters that can seriously harm people or general welfare. Within these areas the code lists specific products that may be used and the practices by which they may be arranged and installed. With products and practices that appear in the building code, there is usually a little more that must be done to gain building department acceptance.

By and large, the products discussed in the building code are those that have been used widely for many years. The code lays out the rules for their use, following traditional, proven practice. The building department rarely questions the use of one of these products if it follows the rules.

However, newer products or unusual uses of established products may not be so readily accepted. The building department may seek more information before allowing construction.

For a product or use that is not in the code, a second resort is to present the building department with an *evaluation report*. This may also be called a *research report*. An evaluation report is an official document prepared by a model code agency, such as the ICC in the United States. In Canada the Canadian Construction Materials Centre (CCMC) prepares the reports. The report discusses exactly how the product can be used in buildings so that it will be in compliance with the rules of the model code (the IRC or IBC or NBC). The report generally covers only some possible uses of the product in certain types of buildings. It does not give blanket approval for use by any method in any building. With an evaluation report, the building department may accept the use of the product. However, they are not required to. If they still have doubts, they may require more supporting information.

Note also that only some products not listed in the code have an evaluation report. Sellers of the product may apply for one. However, they must pay to have the necessary testing and analysis done. They must also pay the model code agency to review this information and write the report. It may be many years before the sellers decide to pursue a report, come up with the money, get all the work done, and receive their report. In the meantime, they will likely sell the product but provide no report.

When no evaluation report is available or the building department is not satisfied with it, the next resort is often to provide the department with *general engineering information* from the seller of the product. Before they apply for an evaluation report, many companies pay testing organizations and engineers to analyze their product and prepare guides that explain how to use it properly. A user of the product can present these to the building department. This may convince the department that the intended use of the product is sound. However, this can be tricky. Many building departments do not feel qualified to evaluate detailed engineering themselves. They may therefore not accept general engineering information from a product supplier.

Perhaps the most powerful support for a product or practice not covered in the building codes is *custom engineering*. This is analysis of the exact product to be used and the exact plans and methods for use in a specific building by a licensed engineer. The engineer studies the plans and the available information on the product and writes a list of rules for exactly how the work is to be done so that it is in conformance with the intent of the building codes and sound practice. In a way, the engineer vouches for the product and how it will be used. The engineer signifies this approval by putting his or her professional stamp and signature on the documents he or she prepares. For this reason, we often say that a particular set of plans "have an *engineer's stamp*."

If the product and the plans for using it are sound, it is almost always possible to get an engineer's stamp. It is also likely that the building department will accept it. With a trained, licensed engineer taking professional responsibility for the plans, the building department is likely to be confident with them.

The major disadvantages of getting an engineer's stamp are the time and cost. On a small building, getting an engineer's stamp for simply the structural elements of the building can cost several thousand dollars and require weeks to prepare. During this time the designers will need to be in close touch with the engineer and may have to modify their plans to meet the engineer's requirements.

Note that the use of unusual products and practices is not the only factor that may cause the building department to ask for extra documents or engineering. The type of building also has great influence. Most jurisdictions require no engineering for a single-family house that adheres to the rules of the building code. But for very large buildings, nearly all departments require thorough, stamped engineering for all important aspects of the building. Even though the building design may conform to the code, the designers must analyze and stamp the plans to verify that they have determined that the plans meet the details of the code, and, where they do not, they follow the intent of the code and sound practice.

Small multi-family and commercial buildings are an in-between case. If they are simple, the building department may not require anything but compliance with the code. However, the larger and more complex the building, the more documents and engineering the department is likely to require.

In sum, the longer the history and the more thorough the documents on a product or practice, the less work is likely to be involved in gaining building department acceptance of it. At the same time, the larger and more complex the building, the more engineering and documentation the department will likely require, even for parts of the building that are well documented.

References and standards

The major building codes frequently refer the reader to other documents for some of their requirements. This is common when there are other organizations that write detailed recommendations for the use of particular products. These are contained in documents called *standards*. Rather than write their own requirements, the building codes state that construction must follow the recommendations given in a certain standard.

The organizations that produce these documents are often called *standards-setting organizations*. They are nonprofit organizations that bring together experts on particular products. These experts write a standard for a product or system, which is a list of the properties (also called *specifications*) they judge that others will want it to meet. In the United States, most of the standards for building products are written by the American Society for Testing and Materials (ASTM). The Canadian equivalent is the Canadian Construction Materials Centre (CCMC). More information is available from these organizations' web sites, www.astm.org and irc.nrc-cnrc.gc.ca/ccmc.

Building codes can require a single product to meet a whole set of important specifications by simply requiring them to conform to the relevant ASTM or CCMC standard. This is also useful shorthand for manufacturers who want to display the specifications. American Society for Testing and Materials and CCMC standards are assigned identification codes as well as descriptive titles. For example, an ASTM standard widely used for concrete masonry has the code ASTM C90 and the title "Standard Specification for Loadbearing Concrete Masonry Units." It requires masonry to meet several different specific requirements.

ACI 318

A very important standard in concrete construction is from the American Concrete Institute and bears the identifying number ACI 318. It is titled "Building Code Requirements for Reinforced Concrete." This standard includes specifications for the materials to use and how to combine them to create building assemblies made of reinforced concrete. The International Building Code in the

United States refers to ACI 318 for most of its rules for constructing structural parts of buildings with concrete. This includes primarily the exterior walls, floors, and roofs. More information is available from the web site of the ACI, www.aci.org.

The International Residential Code also has separate rules from other sources.

Connections

Most of the products and systems of this book connect to other components of the building. In some cases, the methods and details of making these connections are important to understand. Builders and designers need to know how they will integrate with other products currently used.

Further background information on commonly used connections is presented the in the background chapters of Parts 2–6 of the book.

Liaison with architects and engineers

Building professionals are more familiar and comfortable with some construction products than others. They are inclined to accept familiar products readily and work with them without extra support. However, in the case of products that are unfamiliar, they may request additional information. It may be necessary to spend time discussing the products and their use.

For any given product, it is difficult to predict how much of this liaison work may be required. However, this is important information for contractors interested in trying a product they have never used before.

It is possible to give a rough indication of the liaison work that might be necessary. This book summarizes the experience of other contractors using the products in question. It also describes the support materials for building professionals that are available.

Training

For most of the products and systems described in this book, training is available for new installers. This can be valuable for those adopting the products for the first time. The book therefore includes information on this training.

Durability and maintenance

One of the major advantages of most concrete construction products is that they are more durable than alternative products. In most cases, they require less routine maintenance to remain functional and attractive. They also suffer problems that require repairs less often.

The actual recommended maintenance and the historical repair records of these products are clearly important to prospective users. The book therefore presents information on these topics for each product or system.

Energy efficiency

Wall systems, roof systems, and sometimes floor systems influence the amount of fuel a building uses for heating and cooling. A major selling point of many concrete systems is that they can reduce this fuel load. It is useful to understand how and how much.

Exterior walls and roofs influence fuel consumption primarily by slowing the movement of heat between the interior of the home and the outdoors. In fact, it is these walls and the roof, in combination with the windows and exterior doors, that separate the indoors from the outside elements. For this reason, these are, as a group, sometimes referred to as the *envelope* of the building. The walls and roof alone, which make up the structural part of the envelope, are sometimes called the *structural shell* or just the *shell* (Figure 4-3).

In general, the more the envelope slows the movement of heat from inside to outside and vice versa, the less fuel is needed to heat and cool the building. In cold weather, a furnace heats the interior air to a comfortable temperature. However, the heat gradually escapes to the cold outdoors, so periodically the furnace must fire up again and add more heat to maintain the indoor temperature. In hot weather, air conditioning removes heat from the interior, cooling it. However, heat from the outdoors enters and the air conditioner must run periodically to remove it and keep things cool inside.

The more slowly heat moves outdoors in cold weather and indoors in hot weather, the lower the heat loss or gain, and the less the furnace or air conditioner must run. The ability of the envelope to slow the movement of heat has a direct impact on the building's energy consumption.

There are actually several ways in which heat can move from one place to another. Most experts estimate that the ones that are most important for heat movement through building envelopes are *conduction* and *air infiltration*.

4-3 *The exterior envelope (left) and the exterior shell (right) of a building.*

Conduction and R-value

Conduction is the transfer of heat through direct contact. A hot coffee mug warms the hands that are holding it because the heat *conducts* from the hot coffee to the mug and from the mug to the hands. Similarly, in the winter, heated indoor air warms the walls. In turn these warmed walls transfer heat to the cold outdoor air by conduction. In the summer the process is reversed, and the heat outdoors conducts inside.

How fast heat conducts through a wall or roof depends partly on what that wall or roof is made of. Other things being equal, certain materials slow the passage of heat by conduction more than others.

In the United States, the accepted measure of how well a material slows the passage of heat by conduction is *R-value*. The exact definition of R-value is complex. A material has an R-value of exactly 1 if one British thermal unit (or 1 BTU, which is an amount of heat) passes through a section of the material that is one square foot in area in an hour, when the temperature on the "hot" side is 71 degrees Fahrenheit and the temperature on the "cold" side is 70 degrees Fahrenheit. A material that allows heat to pass at only half this rate has an R-value of 2. One that allows heat to pass at only one-third this rate has an R-value of 3, and so on. The higher the R-value, the less rapidly heat transfers.

The R-value of a single material can be tested accurately in a laboratory. In fact, in the U.S. the Federal Trade Commission requires that companies that claim a specific R-value for their materials have it tested by an independent laboratory. If they cannot prove their R-value in this way, they may leave themselves open to fines for false advertising.

The situation is much more complicated for assemblies of multiple materials. A good example is a standard wood frame wall. These are constructed with 2 × 4 or 2 × 6 lumber. The fiberglass insulation installed in most of these walls is 3½″ or 5½″ thick. The thinner insulation is tested to be R-11 or R-13, and manufacturers' markings on it confirm this. The insulation in the thicker 2 × 6 walls is usually R-19.

However, it is rare for an insulated 2 × 4 wall to have an R-value of R-11 or R-13. It is also rare for an insulated 2 × 6 wall to have an R-value of R-19. The reason is that only about 75 percent of the wall is insulated. The other 25 percent of the area of the wall has only wood running through it from front to back. This is where studs, plates, headers, and other wood members are located. Wood does not have as high an R-value as fiberglass insulation does. Therefore heat transfers through the wooden portions of the wall faster than it does through the parts filled with fiberglass. Because of these "breaks" in the insulation, heat transfers through the wall faster than it would transfer through a wall filled everywhere with fiberglass.

A typical 2 × 4 wood frame wall with R-11 insulation that is tested in its entirety is typically found to have an R-value of about 9. Similarly, other frame walls are found to have R-values that are lower than the R-value of their insulation alone.

There is no regulation that wall assemblies, like walls and roofs, must be tested for R-value. Their manufacturers are not generally required to prove their R-value claims. Some will have their assemblies tested and can show the results. However, most estimate these values instead of testing them. There are many ways to make R-value estimates, and they can produce significantly different numbers.

This book attempts to help the reader sort out R-value claims in the descriptions of the various products. Knowing a few rules will help understand these discussions.

The R-value of an assembly is influenced by the type of its insulation, the thickness of the insulation, the breaks in that insulation, and the material located in the breaks.

Type of insulation. Different materials have different R-values. Table 4-2 lists approximate R-values for various insulation and building materials commonly used in construction. An inch thickness of most fiberglass insulation has an R-value of about 3.2 to 3.8, depending on the exact grade of fiberglass used. An inch of plastic foam has an R-value of about 4 to 6, depending on the foam. In any assembly, the higher the R-value of the insulation used, the higher the assembly's total R-value.

Insulation thickness. R-values are *additive*. This means that if a one-inch layer of foam has an R-value of 4, a two-inch layer has an R-value of about 8, a three-inch layer has an R-value of 12, and so on. If there is more than one layer of material in the assembly, their R-values can be added together to get the R-value of the total assembly. For example, suppose a wall had a one-inch layer of foam with an R-value of 4, an eight-inch layer of concrete with an R-value of 2, and a two-inch layer of foam with an R-value of 8. The total R-value of this assembly would be 14.

Breaks in the insulation. As in the example of the wood frame wall, the effectiveness of insulation is reduced if there are other materials cutting through it. You can view these breaks as "leaks" in the insulation layer. The insulation slows the transfer of heat, but a considerable amount of heat may still be able to travel through the breaks. How serious the leaks are depends on their size and their material. If the material breaking through the insulation has a fairly high

Table 4-2. Approximate R-values of Common Construction Materials.

Material	Approximate R-value per Inch Thickness	Factors Influencing R-value
Fiberglass insulation	3.2–3.8	Density of the material
Plastic foam	4–6	Variety of plastic Density of the material
Wood	0.75–1.25	Species
Solid plastic	0.75–1.25	Variety of plastic
Steel	0.002–0.005	Grade of steel

R-value, less heat will leak through than if the material had a lower R-value. Table 4-2 shows R-values for materials that are often used in wall and roof assemblies.

Air infiltration

Regardless of the R-value of the envelope, the energy bill can still be high if a great deal of air leaks through the envelope. The passing of air through the envelope from outdoors to indoors is called *air infiltration*. In cold weather, incoming air will lower the indoor temperature, requiring more heat to keep the indoor temperature at the desired level. In hot weather, warm air will come in, requiring more cooling.

The amount of heating or cooling energy consumed to counter the effects of air infiltration can be high. Some engineering estimates put it at 20 percent to 40 percent of the total heating and cooling load for a typical small building in the United States.

The more airtight the envelope, the lower the air infiltration and the lower the total fuel bill. Unfortunately, there are no common standard tests for the amount of air that a specific wall or roof allows through.

There are, however, some tests for a whole building. These tests measure how quickly air moves into or out of the entire building under certain air pressure conditions. Past tests have shown that small buildings such as houses constructed of wood frame, under common conditions, exhibit a rate of air infiltration equal to about 0.5 *air changes per hour* (ACH). A building is said to have 1.0 air changes per hour, if each hour an amount of outdoor air passes into the house that is equal to the total volume of air contained in the house. In a sense, that full amount of air contained in the house is exchanged for outside air. Similarly, a building with 0.5 ACH allows a volume of air to come inside each hour that is equal to half the amount of air contained in the building.

Thermal mass

Thermal mass is the ability of a material to hold heat. More specifically, it is a material's ability to absorb heat without rising greatly in temperature and give off heat without falling greatly in temperature. A hundred years ago people warmed their toes in their beds at night with grapefruit-sized stones they had heated up in the fireplace. It might take an hour to get the stone warm, but after that it would stay that way for hours. Today we would say that stone has high thermal mass.

Thermal mass is important for building energy efficiency because it can affect the amount of heating and cooling required. It does so without exactly affecting the rate at which heat passes through the building envelope. For that reason it can provide extra energy savings to a building, on top of what results from high insulation and airtightness.

This is easiest to understand for a building in a temperate climate where the outdoor temperature fluctuates widely. If the daytime temperature is 85 degrees and the nighttime temperature is 55, buildings might require cooling during the day and heating at night. But if the building is constructed of materials with a high thermal mass, during the peak heat of the day these materials will absorb some of the heat without getting much hotter themselves. This will help keep the interior cool. When the temperature falls at night, these materials will release heat to help warm the indoor air. The next day the cycle will repeat. The result is that less cooling is necessary during the day, less heating at night, and there is a total energy savings.

According to engineers, there is even some savings in periods of constant cold or constant warm temperatures. The thermal mass helps to reduce the temperature extremes, and this reduces the total heating and cooling energy required.

Total energy savings possible are difficult to predict. Some engineering simulations suggest that annual savings of at least 4 percent to 8 percent are possible when the walls of a small building are constructed with materials having high thermal mass. The total saved will be more in temperate climates and less in climates that tend to be consistently hot or consistently cold.

Sustainability

The *environmental sustainability* of construction refers to its impact on the global environment. Construction is considered to be highly sustainable if the resources it consumes and the waste products it produces will not substantially disrupt or exhaust important parts of the environment over the long term.

Interest in the sustainability of construction has grown rapidly in recent years. Many buyers and designers are requesting that buildings be constructed with methods that have a minimal impact on the environment, and with materials that are available in long-term abundance. There are now scales that promise to measure the sustainability of a building, and some buyers are requiring their projects to achieve specific levels on these scales. It is important for someone interested in using concrete and concrete products to understand how they affect building sustainability.

Unfortunately, sustainability of a building or a product is difficult to measure. It is even difficult to describe briefly. Different aspects of a building or its construction have different impacts on the environment that are hard to compare to one another. However, there are some categories of sustainability and methods for analyzing it that are widely used.

Design and life cycle

There is widespread agreement that the design of a project is important to determining its sustainability. Design typically determines the materials that will be used and how. This in turn determines where the materials will come from and how gathering them will affect the environment, how the building will operate, and the likely lifespan of the building.

All of these factors come into play in the total impact of the building over its lifespan. Experts consider resources consumed and waste generated not only in the construction phase but in the ongoing operations of the building as well. This type of total-life evaluation of a building is referred to as a *life-cycle assessment.*

Some types of environmental impacts may occur only once. These are relatively easy to analyze. Others accumulate over the life of a building and require a life cycle assessment.

An important example is the consumption of energy. The energy consumed in construction is considered to be the source of some of its major environmental impacts. Energy consumption is also something that occurs from before the project begins until the building is no longer used. This includes the activities used to create the building, from the energy used to harvest raw materials for the building and to process them, to transporting the materials and parts to site, and to the assembly of all pieces into the finished building. This energy used to create the building is sometimes called the *embodied energy* of the building. But total energy also includes the fuel consumption to operate the building over its entire life. A good life cycle assessment takes into account the energy consumed in all of these activities.

The same general principle applies to any material. What is important is not solely how much of a material is consumed or how much waste is generated at one point in time, but over the entire life of the building.

Life-cycle analysis is often particularly important to understand with concrete products. In the case of energy, buildings constructed primarily of concrete usually have a higher embodied energy than those constructed with light framing systems. However, many modern concrete wall and roof systems create a building that uses significantly less energy during its operation. The savings are often estimated to exceed the extra embodied energy of the building after a few years.

Life-cycle assessment also helps account for the effects of concrete's durability. Creation of a building with concrete sometimes involves more total material than buildings constructed with other materials. However, concrete buildings often last longer. It may require, say, twenty percent more material to create a highly durable building than a less durable building. However, suppose it lasts twice as long as the less durable building. In that case the more durable building would require fewer materials to build than the two less durable buildings that would have to be constructed to take its place. Because of concrete's durability, it is often necessary to perform this type of analysis when determining its life cycle materials consumption.

Owner environment

Analyses of the sustainability of a building generally also take into account the health, safety, and welfare of the people occupying the building. This is called the *owner environment.*

A major determinant of the quality of the owner environment is considered to be the presence of *indoor air contaminants*. These include gasses that may escape from building products (including *volatile organic compounds*, also called VOCs), mold, dander, dust mites, and major combustion pollutants.

Concrete is one of the most inert of all construction materials. It releases no significant amounts of any known harmful gasses. Because it is inorganic it provides no nutrients to support the development of mold or mites. It does not burn or produce significant combustion products at the temperatures found in typical building fires.

Other properties of concrete affecting the owner environment are its thermal mass, fire resistance, and *sound attenuation*. Thermal mass is the ability of a material to absorb and retain heat. Concrete does this more than many other construction materials. If substantial portions of the building consist of concrete, it can have the effect of leveling out swings in the indoor temperature, increasing comfort.

Fire resistance adds to occupant safety. Because portions of the building made of concrete generally do not burn, they are less likely to collapse during fire. Note, however, that other hazards of fire, such as from combustion smoke of other materials, will still be present. Construction of large parts of a house out of concrete by no means eliminates fire risk. Occupants are still advised to take recommended fire precautions.

Concrete attenuates (reduces) sound that passes through it. Less of the sound occurring outdoors passes through concrete walls and roofs than through light frame. This can be especially beneficial in areas of high noise.

Local and global environment

The other factors taken into account in a sustainability analysis are sometimes classified under *local environment* and *global environment*. Different speakers use these terms in different ways. However, they are usually considered to include the building's impacts on the outdoor environment. These include effects of extracting and processing the materials, transporting them to the site, assembling them into the building, and operating that one building over its life. Some of these were described previously in the discussion of life cycle assessment.

Gathering materials

The effect of harvesting the raw materials for building is considered one of the greatest environmental impacts of construction. The materials for steel must be dug from the ground and processed from ore into products like beams and steel studs. For cement we dig up the geological elements calcium, silicon, aluminum, iron and gypsum. These are mixed with water and aggregates, which are also mined, to form concrete.

Wood harvesting seems quite simple, just cut down trees. But, this can be complex. Only certain methods of logging are considered "sustainable." Sustainable logging does very little environmental damage, and that can be recovered. If non-sustainable logging methods are used, however, it can seriously disrupt local, regional, or global ecosystems. In some cases these ecosystems may never recover.

The impact of mining iron ore may be extensive. It is most often found in mountain environments. These may be seriously affected by the land degradation from open pits, waste tailings, and large dump pits of unusable material. Currently, in some countries, there are attempts being made to reforest old mining land. Other regions are going without any serious attempts to maintain a sustainable use of the land.

The major ingredients of concrete are also mined from the earth. The cement comes mostly from shale or limestone. These materials are abundant in many places, and so they are mined locally in many regions of the world. The environmental impacts include dust, the effects of excavating quarries, and energy requirements for gathering and transporting materials. These are serious issues, but the efficiency of the mining operations in North America is increasing. It is also important to note that cement makes up only about ten percent of the final concrete. Eighty percent of a typical concrete mix consists of aggregates.

The sand and stone aggregates of concrete can be mined without deep excavation compared to the other minerals. Much of the removed material is used, unlike in the mining of some metal ores. With some ores, only a fraction of the excavated material is actually used. With aggregates there are no emissions from smelting or other mining processes. The material has little need for alteration or cleansing before going to processing.

Processing materials

The major impacts of processing raw materials result either from the fuel consumption of the processing or the waste creation.

Producing cement consumes significant amounts of energy and waste. Making one ton requires about 3 million BTU of power and nearly two tons of raw materials. It releases about one ton of carbon dioxide (CO_2), and some other emissions. But as noted previously, cement comprises only about ten percent of concrete. The aggregates, which make up about eighty percent of concrete, involve little processing after they are gathered.

Steel processing also involves significant energy and waste. Blast furnaces melt off the iron from other impurities. It is then oxygenated at high heat and processed further to remove any remaining impurities and add further chemical additives.

Wood processing involves less energy and waste. The trees are felled. The lumber is milled and kiln dried. Wood is also often transported long distances to reach the customer. Yet, this still requires less energy than the processing of similar amounts of concrete and steel. Processing wood produces waste in the form of unused parts of the tree. However, if sustainable logging practices are used, these are usually small and much of the waste can be used in the production of other goods such as paper and fiberboard.

The wood building typically lasts a shorter time than the concrete building. So the lower processing requirements of wood must be weighed against the need to perform that processing more often for wood buildings than for concrete buildings.

Note also that the waste and emissions from cement production in North America is declining. Over the last 10 years, the manufacturers have carried out major programs to make their processes more efficient and use more recycled materials. Seventy-five percent of cement kiln dust is now recycled, amounting to nearly 8 million tons per year. It goes back into the production kiln, reducing the need for raw materials. The industry has also adopted a program of using recycled waste materials from other industries; this includes 11.4 million metric tons of fly ash in 2001 from coal-burning industry and about 3 million metric tons of recycled slag waste from the steel industry. Even used concrete is sometimes recycled. It is crushed and used as an aggregate for new concrete (Figure 4-4).

4-4 *A modern cement plant.* Portland Cement Association

Building energy consumption

Buildings constructed with modern concrete systems are frequently more energy-efficient than those constructed with wood or steel frame. They use less energy for heating and cooling. Since energy consumption also releases CO_2, the concrete building is also responsible for less of that pollutant. The extent of the savings depends on the exact concrete components used and other factors. However, estimates suggest that in many situations the energy and carbon dioxide savings of heating and cooling a concrete building exceed the additional energy and CO_2 involved in producing the materials after a few years of building operation.

Leadership in energy and environmental design

Leadership in energy and environmental design (LEED) is a method of measuring the sustainability of a building. It was created in 2001 by the National Green Building Council (USGBC), a nonprofit organization established for this purpose. According to the USGBC, "Through LEED's use as a design guideline and third-party certification tool, it aims to improve occupant well-being, environmental performance, and economic returns of buildings using established and innovative practices, standards, and technologies."

The LEED rating system has become extremely popular among building designers and owners. They now frequently require that their buildings attain a certain level on the LEED scale so that they can feel comfortable that they are "green" buildings. In 2004, there were over a thousand buildings worldwide that achieved a LEED certification, and the number is increasing rapidly.

The LEED system is currently designed only for rating new commercial buildings. However, the USGBC intends to release new scales for rating homes and renovation projects as well.

The system awards points to a building based on environmentally responsible construction methods, materials, or architectural site and building designs. There are sixty-nine possible points to be awarded. To receive a LEED certification the building must meet certain basic requirements, which are spread over several diverse areas. All other points are optional, but the building must also achieve a minimum of twenty-six total points. Buildings that attain more than twenty-six points may receive higher "levels" of certification. Table 4-3 lists the levels of LEED certification.

Table 4-3. Levels of LEED Certification.

Points Achieved	LEED Certification Level Awarded
26–32	Certified
33–38	Silver
39–51	Gold
52–69	Platinum

Source: U.S. Green Building Council

Awarding LEED points is based partly on the judgment of the officials of LEED, who review the application materials. However, experience has shown that the use of concrete or concrete products can help achieve several of the points available on the LEED scale. Some of these points are as follows:

Credit 1 Optimize Energy Performance   1–10 points
This is a broad category that awards points based on the overall energy efficiency of the project. The use of concrete walls, floors, and roof might achieve several points. The number will depend mostly on the R-value, air tightness, and thermal mass of the products used, as well as how much of the structure is built with them.

Credit 4.1 Recycled Content (5%)   1 point
Credit 4.2 Recycled Content (10%)   1 point
Points are awarded based on the amount of recycled materials used in the building. Two separate points are available, depending on the percentage of the material that is recycled. Concrete can contribute to this content if it includes such ingredients as fly ash, blast furnace slag, or recycled aggregate.

Credit 5.1 Local/Regional Materials (20% Manufactured Locally)   1 point
Credit 5.2 Local/Regional Materials (50% Harvested Locally)   1 point
Up to two points are also awarded based on how close the production of the building materials is to the job site, which affects how much energy must be used for transportation. These two credits often apply to concrete use. The aggregates (nearly 80% of the mix) are virtually always local, and in many cases so is the cement.

Credit 6.1 Stormwater Management   1 point
This could be awarded for use of pervious concrete flatwork or certain pavers, where storm water passes through the pavement into the ground.

Credit 7.1 Heat Island Effect (Non-roof)   1 point
This could be awarded for two types of concrete use. It might be awarded for use of pavers that allow grass to grow through the openings. In warm conditions moisture from the open ground evaporates, keeping the area cool. It might also be awarded based on the use of concrete paving, instead of a dark paving such as asphalt. The lighter concrete paving will reflect more sunlight and heat up less.

Credit 7.2 Heat Island Effect (Roof)   1 point
This could be awarded for the use of a light-colored concrete roof.

PART II

WALL SYSTEMS

5

Background on concrete wall systems

Over the last decade there has been a surge in the use of concrete for the exterior walls of small buildings. There are several different concrete wall systems that have contributed to this growth. They are different from one another, and each one provides contractors and buyers with its own set of advantages. Yet all of them also share many properties. These are properties that result from use of the concrete material itself.

History

Concrete has been used in walls since ancient times. However, during most of history this was merely as the mortar that bonded stones, bricks, or other building units to create a structural wall. In the early 1900s, construction pioneers experimented with walls built completely of concrete. Some of these were cast-in-place with removable forms. Others were built with concrete masonry and mortar.

By World War II a large portion of the small buildings in the United States were constructed with concrete exterior walls. Concrete had effectively replaced the structural brick and stone walls that were popular before then because it was generally faster, less expensive, and more consistent. Nearly all foundation walls were either masonry or cast-in-place with removable forms. Below grade, wood frame was usually not competitive because of the need for high strength plus moisture and insect resistance.

Above grade also, many structural walls were constructed of concrete masonry. Wood frame was more competitive above-grade because of its speed and low cost. However, in larger multifamily and nonresidential buildings, requirements for strength, durability, and low maintenance favored concrete.

In homes, concrete masonry was the favored material in many of the southern states. Concrete fit these areas well. Wood-damaging insects are more active in warm climates. Much of the South is wet, and concrete was less susceptible to moisture damage. The thermal mass of the masonry walls was also useful in moderating the day-night swings in temperature.

The National Concrete Masonry Association estimated that up to 20 percent of the houses built in the U.S. in the late 1940s had concrete above-grade walls. This share was certainly higher for small commercial, institutional, and industrial buildings.

However, concrete's share declined steadily after WWII. This is partly because of constant streamlining of wood frame construction that steadily made it faster and cheaper. In addition, the Energy Crisis of the early 1970s and the sharp rises in energy prices that followed had a large impact. At the time, there were no widely available methods of insulating concrete walls that were as cost-effective as insulating frame walls with fiberglass insulation.

By the early 1990s, only about 3 percent of all houses constructed in the U.S. had concrete walls above-grade. Nearly all of these were in the southern half of Florida. There the advantages of concrete were most important, and the need for a high R-value was lower.

The use of concrete for the walls of commercial buildings appears to have declined as well. However, it did not drop as much because of the need for concrete's properties in larger and heavy-use buildings.

Construction with concrete walls began to increase again in the mid-1990s. One reason for this was the fallout from some major windstorms. Hurricane Andrew and later hurricanes hit the Southeastern states. These caused extensive damage to the buildings constructed with wood frame. The buildings with reinforced concrete walls fared better. As a result, buyers began to favor concrete walls. Building officials also put greater requirements on wood construction to strengthen it, which increased the cost of building with wood.

In addition, several new or modified systems for building with concrete walls became available. The new systems included insulating concrete forms and autoclaved, aerated concrete (AAC). Precast, cast-in-place, and tilt-up systems had long been common on larger buildings, but by the 1990s versions of these systems suited to small buildings had also been developed. Suddenly buyers could choose the concrete wall systems that best met their needs from several different options. Many of these new and improved systems offered the high R-values that are important in many applications.

The Portland Cement Association estimated that in 2003 about 16 percent of all single-family homes were constructed with exterior walls of concrete. Manufacturers also reported significant growth in the use of concrete wall products in multi-family and nonresidential buildings.

Market

Almost every segment of the small building market is now using a significant number of walls made with concrete. Because of the diversity of systems available, there are options that are well suited to almost every use. However, different systems appeal to different segments. Chapters covering the individual systems explain where each one has had its greatest sales.

Advantages

The different wall systems have different sets of advantages to building owners and contractors. However, a few owner advantages are common to all the concrete systems. By and large, these come from the nature of the concrete material itself.

All concrete wall systems are believed to have high resistance to natural disasters. This holds particularly for high winds, including hurricanes and tornadoes. Engineering calculations and surveys of areas that have had great damage from wind indicate that reinforced concrete walls survive the destruction well. Where extra strength is necessary to meet high requirements, it can be obtained by increasing the amount of steel reinforcement in the walls. This is usually an easy and inexpensive adjustment. Therefore, the walls can be readily changed to adjust their strength as desired (Figure 5-1).

The same principles appear to hold true for resistance to earthquakes and floods. Since these events are less common, there are fewer available statistics. Concrete's resistance to earthquake damage depends heavily on how much steel reinforcing is in the wall. Steel provides the ductility needed to resist the shaking motions. For that reason, the reinforcement requirements on concrete are highest in high seismic zones. However, the additional rebar is not particularly expensive. Experience with modern reinforced concrete buildings in recent Western earthquakes has shown the buildings to hold up well.

All of the concrete wall systems have good resistance to fire. Fire is unlikely to compromise the concrete structure significantly or pass through concrete walls readily. However, it is always important to remember that much of the damage and loss of life from fire results from interior fires that consume the building's contents, before reaching the exterior walls. Therefore all the standard fire precautions are important, even in a building with concrete walls.

Regardless of the concrete wall system used, the wall structure is also highly resistant to insects, rot, mold, and mildew.

Virtually all concrete walls also provide some energy efficiency because of their air tightness and thermal mass. R-values of the wall systems vary, but for most the R-value can be adjusted from very low to very high, depending on the need.

Concrete walls reduce the transmission of sound from outdoors to indoors more than conventional frame walls do. The amount of sound that penetrates

5-1 *Foundation of a wood frame home blown away in a tornado (above), and the concrete home next door that survived (below).* Reward Wall Systems, Inc.

from such outdoor sources as traffic and lawnmowers can be significantly less. This will vary with the wall system. Tests show that conventional wood frame walls have a *sound transmission class* (STC) of about thirty-six. Tests of concrete walls show them on average to have an STC of about ten points higher, in the range of forty-four to fifty. This indicates that the concrete wall will allow only about one-third as much sound energy to penetrate as will the frame wall. This is clearly noticeable.

The installation of the available systems is so different that it is impossible to generalize about the advantages to contractors. The contractor needs to understand the different systems and pick the ones that will be best for each particular operation.

The wall assembly

The wall assemblies constructed with the different systems are sharply different from one another. Yet a few basic things are common. All are primarily concrete. All of them normally include steel reinforcing bar. The amount and placement of rebar can generally be varied to adjust the strength of the wall as needed. In many systems, the thickness of the concrete can also be adjusted to meet strength requirements.

Foundation connection

The wall systems connect to the foundation beneath them in different ways. The most common method is with short rebar, sometimes called *dowels.* The bottom ends of the dowels are embedded in the foundation. The upper ends extend out of the foundation vertically upward. During construction, the upper ends of the dowels are embedded in the walls.

Insulation and electrical lines

Most of the wall systems include at least one layer of plastic foam for insulation. In some of them the thickness of the insulation may be varied to increase or decrease the R-value of the wall as desired.

With virtually any concrete wall system, it is also possible to install furring or studding to the inside face of the wall. Fiberglass or other insulation typically goes into the cavities. Wallboard is installed over the studding for the interior finish. With some systems this is an alternative method of insulating, instead of foam. With others it is the usual method.

Electrical wiring may be installed in a variety of ways. In wall systems with foam, it can usually go into chases cut into the foam. In systems with studding or furring inside, it can go into the spaces between studs or furring strips, much as it does in frame construction. As another option, in some systems it can go in conduit that is cast into the concrete (Figures 5-2 and 5-3).

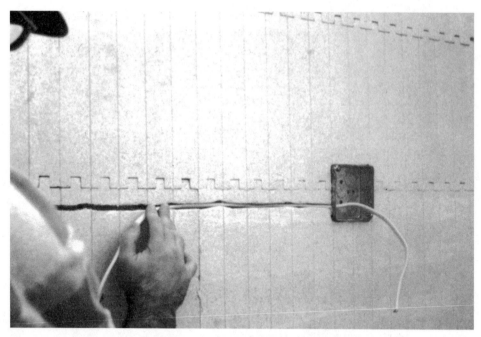

5-2 *Electrical cable installed in a chase cut into insulating foam.* Portland Cement
Association

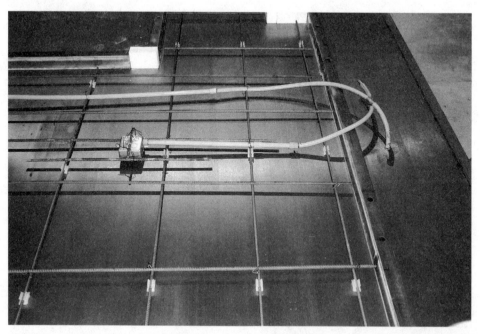

5-3 *Electrical conduit in position before placing concrete for a precast wall panel.*
Dukane Precast, Inc.

Doors and windows

Openings with doors and windows are created in various ways. One method common to several wall systems is to use a *buck*. A buck is a frame inserted into the wall to form the opening. Usually it is constructed of dimensional lumber. Since it will be in contact with concrete, most building codes require that the lumber be moisture-resistant material or be covered with a water-resistant coating or membrane (Figures 5-4 and 5-5).

Bucks are frequently left in place to provide an attachment point for the door or window. Bucks left permanently in position are usually connected to the wall with some form of anchor so that they cannot shift after the concrete cures.

Another approach is to make a frame that is removed after the concrete cures. Some refer to this as a *blockout* since it blocks the concrete from entering the space of the opening. The window or door is then fastened directly to the concrete. This is common in nonresidential and larger buildings. The windows and doors are usually designed to be attached to the wall with concrete fasteners run through holes in the frames.

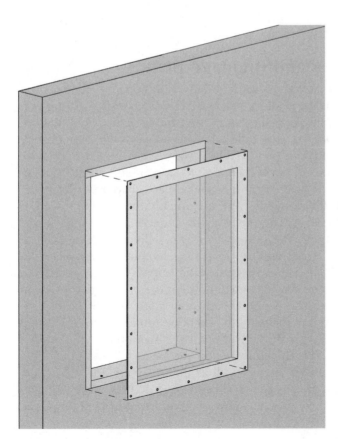

5-4 *Attaching windows to an opening formed with a permanent buck.*

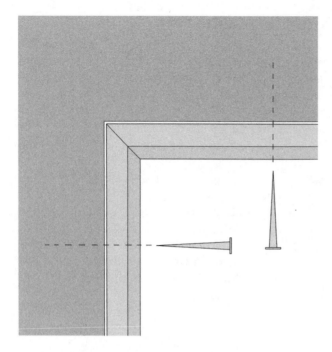

5-5 *Attaching a screw-through window frame to an opening formed in a concrete wall with a blockout.*

Wythes and drainage planes

An important wall option that comes up frequently with concrete construction is the number of wythes in the wall. A *single-wythe* wall consists of one vertical wall assembly. For example, a standard wood frame wall would be considered single wythe. A *double-wythe* wall has two vertical walls with a cavity of perhaps a few inches between them (Figure 5-6).

A single-wythe wall needs to be tight enough to stop air and water from entering from the outdoors. For this reason it is sometimes called a *barrier wall.*

The inner wythe of a double-wythe wall is normally the structural wall. It holds up the floors and the roof. The outer wythe is decorative. It also serves as the outer shell of the cavity. The outer wythe is usually constructed out of brick or other masonry. It is designed to stop the elements just as a single-wythe wall is. However, if moisture penetrates the outer wythe, it enters the cavity and generally drains out through small holes at the bottom before it can enter the building's interior. The holes, called *weep holes,* are in the face of the outer wythe at the bottom of the wall. In addition, humidity can be flushed out by circulation of outside air through the cavity. A double-wythe wall is sometimes called a *cavity wall.* The construction can be modified to include a layer of foam insulation in the cavity, as well as the air space.

Traditionally, double-wythe walls have been regarded as a high quality form of construction. The cavity prevents unwanted moisture from entering the

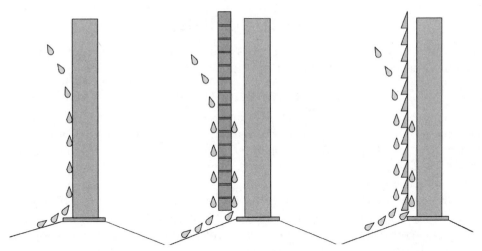

5-6 *Flows of rain against a single-wythe wall (left), double-wythe wall (center), and drainage plane (right).*

building. The option for a nearly unbroken layer of insulation in the cavity can provide high energy efficiency.

Many of the newer concrete wall systems have any of the benefits of double-wythe construction in a single wythe. They are designed to create effective barriers against water and wind because they are cast as a nearly continuous material. They may also have continuous insulation layers.

In some cases single-wythe walls are outfitted with a thin cavity. This is typically done by furring the outside of the wall and then affixing a lightweight, rigid exterior finish to the furring. If this is detailed correctly, water can drain out of the cavity, and air can circulate through it. Such an air gap built into the wall assembly is sometimes referred to as a *drainage plane* or *rain screen.*

Code and regulatory status

Concrete walls have over a century of experience and are backed by extensive engineering. Existing building codes include full sections on wall construction using many of the concrete systems. Building department acceptance in these cases is usually straightforward.

Many of the newer systems use the same principles of concrete construction and engineering. However, since they are new, they are not always covered as thoroughly in the building codes and engineering texts. Building departments may request documents or engineering to support the use of these. The situation with each system is somewhat different, and is therefore covered in the separate chapters on the individual systems.

Connections

The major connections made to exterior walls are to floors and roofs, interior fixtures and finishes, and exterior finishes.

Floors and roofs

The connections of floor decks and roofs to exterior walls are generally considered structural. These parts of the building bear heavy loads, such as wind against the building, the weight of snow, and the weight of people and furniture. The connections between these elements must transfer these loads from one element to the other without failing.

In a small building, the floor or roof is often some type of light frame, either wood or light-gauge steel. In this case the connection is usually by means of some type of steel connector that is embedded into the concrete of the walls. It is later fastened to the framing.

One common connector to framing is the *anchor bolt*. This may also be called a *J-bolt*. It is a steel rod that is bent at a right angle at one end to resemble the letter "J." The other, straight, end is threaded to receive a nut. The bent end is embedded in the concrete while it is wet. The threaded end extends out of the wall where the connection is to be made. The frame members are then bolted to the exposed anchor bolts.

Anchor bolts come in varying lengths for different purposes. They are also available in different diameters to provide different levels of strength. In small building construction, most are a ½-inch or ⅝-inch in diameter.

Note that there are alternative versions of anchor bolts that do not need to be embedded while the concrete is wet. Instead, holes are drilled in the hard concrete later, and the anchoring end of the concrete is inserted and permanently adhered with epoxy or other construction adhesive. This device is handy when embedded bolts were left out or the when it was easier to determine the proper location of the bolt later.

The most common method of connecting a frame floor to a concrete wall uses a *ledger* connected to the wall with anchor bolts. According to this method, the bolts are installed in the concrete. The ledger is later drilled out at the bolt locations, fitted over the bolts, and screwed tight to the wall with washers and nuts. For a wood floor the ledger is usually 2× lumber. The width of the ledger is chosen to match the depth of the joists to be used. The joists fasten to the ledger with conventional joist hangers. A wooden ledger must usually be water-resistant or have a coating or membrane, since it will be in contact with concrete (Figure 5-7).

When the frame floor is light-gauge steel, the ledger is typically a steel track. The joists fit into the track and attach with screws. Typically one screw goes through the upper flange of the track into the joist, one through the lower flange into the joist, and others may connect the web of the joist to the web of the track with a right angle piece of steel called a *clip angle* (Figure 5-8).

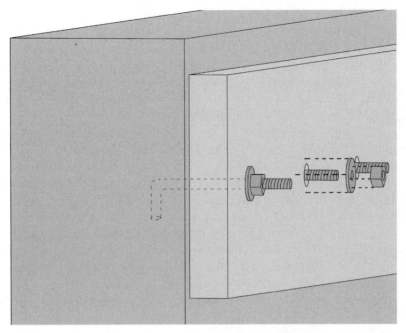

5-7 *A wooden ledger attached to a concrete wall with anchor bolts.*

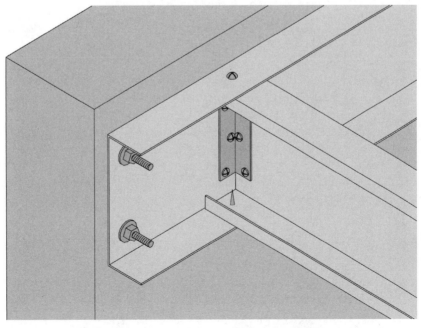

5-8 *Steel floor framing attached to a concrete wall with anchor bolts.*

Also commonly used on frame are various types of steel straps. They are appropriate for connecting a roof down to the top of a wall. These are also embedded in the concrete of the wall, with their ends extending out the top. The free ends usually have holes in them for nailing or screwing the strap to framing.

One of the most common methods of connecting a frame roof to a concrete wall is with a *top plate*. The workers producing the walls are typically responsible for embedding either anchor bolts or *sill straps* into it along the top edge. Sill straps are steel straps with a deformed end that embeds into the concrete and two free ends that extend out of the concrete to wrap around the plate. The framing crew later attaches a plate to the top of the wall with the bolts or straps. They can fasten the roof framing to the plate, just as they would fasten it to the top of a frame wall (Figure 5-9).

The other common roof attachment method is with *hurricane straps*. "Hurricane strap" is the popular name of a variety of steel strap that is designed to wrap directly around roof rafters or trusses. The wall crew embeds these along the top of the wall like sill straps. The framing crew wraps one around each rafter or truss and nails or screws the strap to it (Figure 5-10).

Hurricane straps are so named because of their great strength against the uplift forces of winds. They are the usual method of attaching a roof in some

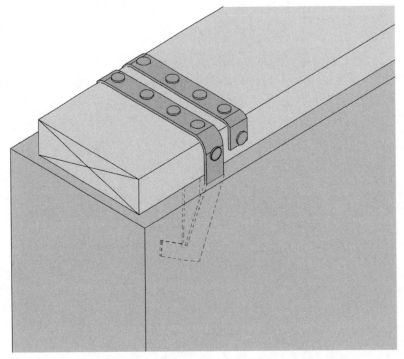

5-9 *A top plate attached to the top of a concrete wall with sill straps.*

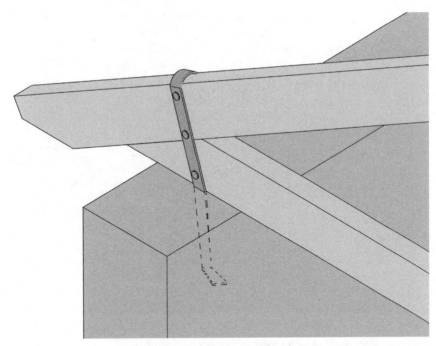

5-10 *Roof truss attached to the top of a concrete wall with hurricane straps.*

high-wind areas. Some anti-disaster organizations recommend them for roof attachment in all areas. However, using them efficiently depends on knowing the position of the roofing members when constructing the walls so that the straps can go in the correct locations. Some contractors find it simpler to install a plate, since that allows attachment of the rafters and trusses anywhere along the wall.

In commercial construction, heavy steel frame floors and roofs are often attached by means of *weld plates* or *bolt plates*. These are heavy-gauge steel plates with deformed steel rods attached to one face. The plates are positioned so that the rods extend into the wall and are cast into the concrete. The opposite face of a weld plate is flush with the surface of the wall. The parts for steel floors or roofs can be welded directly to the plates. Bolt plates usually protrude out from the wall. They have holes in them for bolts to connect to other members (Figure 5-11).

When the floor or roof is constructed of concrete, the connection is usually accomplished with rebar. Bars extend from within the wall to within the floor or roof. A bar must be embedded at no less than some minimum distance into each element, which is set by the engineering requirements. Usually this is about two feet. Common connections of concrete walls to concrete floors and roofs are covered in Part 3 of this book.

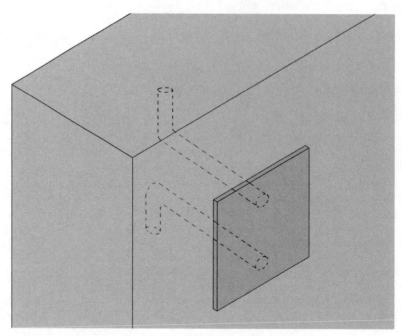

5-11 *Steel weld plate embedded in a concrete wall.*

Interior finishes and fixtures

The interior finish on concrete walls inside a small building is usually conventional wallboard. This may be screwed directly to the wall if the wall contains some sort of surface that can receive a light fastener. Several wall systems include these *fastening surfaces*. These are usually strips of wood, plastic, or light-gauge steel that run along the wall surface.

If the wall is plain concrete, it may not include spots for attaching light fasteners. In this case the wall may be simply plastered or painted, depending on requirements. If wallboard is preferred, it may be attached with glue. It may also be attached to wood studding or furring. Studding amounts to building a frame wall and installing it along the inner face of the concrete wall. It can be used and finished as would be done with a conventional stud wall. Furring consists of shallow strips of wood that are fastened to the face of the concrete. This is usually done with concrete nails. The strips are arranged vertically, with one about every 16 inches. This is similar to the pattern of conventional studs in a wall. Studding or furring also provides a cavity for electrical lines and insulation (Figure 5-12).

Interior fixtures can also attach to the wall system's fastening surfaces, studding, or furring, if one of these is available. They can also be attached directly to the concrete with concrete nails or screws. For very heavy fixtures, designers often plan ahead to embed anchor bolts in the desired location and bolt

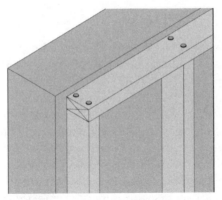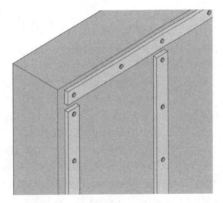

5-12 *Concrete walls outfitted with studding (left) and strapping (right).*

the fixture to the wall. An alternative is to install epoxied bolts in the concrete later, when the fixture location is known exactly.

Exterior finishes

Virtually every popular exterior finish is applied to concrete walls. The methods of attachment used can vary with the type of wall system used. For some systems, certain finishes are easier and more practical to attach than others.

An exterior concrete surface can be troweled directly with stucco products. This can be highly economical. For a stucco finish over frame walls it is usually necessary to install a weather-resistant membrane and a steel or fiberglass mesh. Neither of those is typically required over a concrete surface.

Masonry veneers, such as stone, brick, or architectural block, go up the same way they do on any wall. They are connected to the wall with steel straps that are commonly called *brick ties*. Depending on the wall system used, the ends of the brick ties may be connected to the concrete wall by screwing them to a fastening surface that is part of the system, screwing them with concrete screws directly to the concrete, or embedding them in the concrete. The opposite ends of the brick ties are embedded in the mortar joints of the veneer as it is built.

Exterior finishes that are commonly nailed or screwed to the wall include lapped sidings (vinyl, clapboard, fiber-cement) and various panelized and board products. These can be nailed or screwed to the fastening surfaces of the wall system or to furring straps installed to the face of the wall.

Architects and engineers

Architects and engineers are generally familiar with the traditional concrete wall systems. They may be less familiar with some of the new ones. They will likely need support information and some preparation time to work with unfamiliar systems.

Energy efficiency

The walls of a building have a large influence on its energy efficiency because they make up a large proportion of the total building envelope. In most small buildings, they are typically about half.

All of the mainstream concrete systems are believed to provide some energy savings because of two characteristics of concrete walls. These are their relative air tightness and their thermal mass. Concrete walls allow relatively little air to infiltrate because of how they are constructed. Nearly all of the systems create an almost seamless layer of concrete. This is true regardless of whether they are cast in place, precast, or assembled out of pieces of masonry. Many of these systems have no joints. With others, the joints are effectively sealed with mortar or sealant. Concrete walls also provide a much more massive wall than standard wood frame. This helps to even out temperature fluctuations, shaving energy costs.

The R-values available with concrete wall systems range from low to very high. Users pick the system and parts that best suit the local situation and their energy-saving preferences. When a high R-value system is used, the high insulating value combined with low air infiltration and high thermal mass makes some of the most energy efficient walls that are currently available with established construction methods.

Maintenance and repair

In most cases, there is no routine maintenance and little or no expected repair necessary for concrete wall systems. The concrete structure does not rot or decay, and is not subject to insect damage. It does not support the growth of mold or mildew.

The finishes covering the wall may require maintenance, as they would when applied over any wall surface. This depends mostly on the finish used, not on the concrete behind it.

Wall systems that have a layer of foam or other soft material on one or both wall surfaces must generally be protected. The interior is normally covered with wallboard. This allows little wear to the soft material behind it and durability is high. On the exterior, it may be important to choose a durable finish material. Particularly in situations with high traffic or vehicles or heavy equipment passing close to the building, a soft finish material may allow occasional damage.

Concrete wears well even when exposed to the weather. When a concrete exterior is painted it usually requires repainting less often than wood does. Typical recommendations from manufacturers are to repaint wood siding every five years, but repaint concrete surfaces about every seven. Some systems enable the use of novel finishes cast into the concrete surface itself. These are generally durable as well.

Sustainability

Concrete walls can contribute to obtaining several LEED points for a building.

Concrete walls can help significantly with the energy efficiency of a building, which can lead to the award of up to ten points. At least some of these points may be available for almost any concrete wall system because of its air tightness and high thermal mass. When using a system with very high R-value, the walls may help qualify for more. Note that to achieve the full ten points other key components influencing the total heating and cooling load (roof, windows and doors, heating and cooling equipment, design, etc.) would also need to be energy efficient.

The walls can also help qualify for the recycled materials point. The concrete may contain significant amounts of recycled materials such as fly ash or blast furnace slag.

The concrete or concrete components are almost always produced in the area local to the job site. This helps obtain the point for locally manufactured materials. In addition, the aggregates and sometimes the cement ingredients come from local quarries. These may contribute to the point for ingredients that are locally harvested.

Separate from the LEED credits, building with concrete walls is often cited as a method of avoiding the harvesting of trees.

Aesthetics

Because the core of the wall is covered, it is the finishes put over it that determine its appearance. Most concrete walls are covered with the same finishes used over other structural walls.

On the inside the finish material is usually gypsum wallboard or plaster. In industrial buildings using heavy equipment, the concrete surface may be left unfinished to economize or leave a highly durable surface.

On the outside, common finishes include stucco products, brick, vinyl, stone, and lapped siding products. Designers select the finish that provides the desired appearance, as with most other wall systems.

With the systems that provide an exposed concrete exterior face, it is often possible to create an architectural concrete surface. These surfaces may simulate other materials, such as brick or stone, or have distinctive colors and textures that are only practical with concrete.

Some of the exterior finishes can be installed more quickly and economically onto some of the wall systems than onto others. As a result of this, with some wall systems the finishes that can be most readily applied tend to be favored by designers and owners. The finishes that happen to be most economical vary from one system to another.

Key considerations

Most of the important considerations to keep in mind when building with concrete walls depend on the system used. However, a few apply in many or most cases.

Rest of the building envelope

It is important to remember that many of the benefits of concrete walls depend also on using quality components for the other parts of the building envelope. Energy efficiency, disaster resistance, and sound reduction all depend on how well the concrete walls isolate the interior from the outdoors. But if the roof, windows, and doors do a poor job of this, much of the benefits can be lost. For example, some buildings with concrete walls and light, poorly attached roofs have gone through hurricanes. The walls remained in place, sound, and largely undamaged, but the roof frequently blew off. Wind and rain came in freely from above, destroying most of the contents of the building. Despite the stable walls, the cost and extent of the damage was high. Any occupants would have been at risk.

The same applies to many other benefits of concrete walls. If the roof and windows are leaky and have a low R-value, the building may still have a high fuel bill. The walls may stop sound, but if it can pass freely through light windows and a loosely-constructed roof, traffic and airplanes can still be clearly audible in at least some parts of the building.

Because of this, when a building is designed to gain the benefits of concrete, it is advisable to think through the other parts of the envelope as well. It may be worthwhile to upgrade them so that they too have some of the premium performance benefits of the walls, such as high strength or energy efficiency.

Heating, ventilation, and air conditioning equipment sizing

When a building is constructed with energy-efficient concrete walls and other efficient components in the rest of the envelope, there is a danger that the heating and cooling equipment will be oversized. Many heating, ventilation, and air conditioning (HVAC) contractors working in small buildings estimate the size of the equipment needed from the size of the building. This makes sense if the building envelope is of average quality. However, many concrete wall systems make the envelope much more energy-efficient than average. Therefore, smaller equipment may be sufficient.

Unfortunately, traditional HVAC contractors may not be willing to install smaller equipment. If they are told that the building will be more efficient, they may be skeptical. Their concern is that if they install smaller equipment, and it turns out not to be big enough, they will be blamed. To guard against this possibility, they may install the normal equipment.

There are strong disadvantages to oversized HVAC equipment. It will cost more than need be, robbing the customer of some potential savings. The large equipment will tend to run only for short periods. This is because a large furnace or air conditioning unit will fill the building rapidly with large volumes of heated or cooled air, bringing the interior quickly to the desired temperature. However, short runs of the equipment are usually inefficient and can cause it to wear out more quickly. In addition, in humid climates the air conditioning also serves to remove moisture from the air. If it runs for only short periods it may not dry the interior air enough to be comfortable, even if it is cool enough.

There are ways to prevent the oversizing of equipment. One is to find an HVAC contractor experienced with highly efficient or super-insulated buildings. These contractors will be more likely to size the equipment correctly. Another is to find a contractor that uses software to select the equipment size. Many HVAC programs can now take high R-values, low air infiltration, and high thermal mass into account in their calculations.

There is now inexpensive software available from the Portland Cement Association (PCA) for sizing the HVAC equipment in houses with concrete walls. It is designed to be easy to use. Titled "HVAC Sizing for Concrete Homes," the program is available on the PCA web site, www.concretehomes.com.

This problem tends to be less in larger buildings. Large projects frequently have a mechanical engineer assigned to calculate the heating and cooling loads and select the equipment. The engineer typically does a more detailed analysis that takes many factors, like high R-values, air infiltration, and thermal mass, into account.

Air exchange

Using concrete walls will likely produce a building with lower air infiltration than a building with frame walls. If other parts of the envelope are also tight, the air inside may become stagnant and accumulate indoor air pollutants. There is currently no minimum code requirement for how much air a building allows in from the outdoors. However, the American Society of Heating, Refrigeration, and Air Conditioning Engineers (ASHRAE) is considering recommending that at least some buildings have a rate of at least 0.35 air changes per hour (ACH). Measured air infiltration rates for houses built with concrete wall systems have frequently been close to 0.35 ACH. The addition of a tightly constructed roof would probably reduce the rate below that.

If there is a concern that the building will be too tight, there are several methods of drawing controlled amounts of fresh air into the building through the HVAC system. These can also be designed to draw the air while maintaining energy efficiency. One popular, effective method is to use a heat recovery ventilator (HRV) or energy recovery ventilator (ERV). These can provide high ventilation rates while keeping much of the heating or cooling of the inside air from escaping. It is best to consult with the HVAC contractor about the best way to accomplish this in a particular building. Contractors with more experience in tightly constructed and super-insulated buildings will probably be most helpful. Again, the problem is less of a concern in larger projects where a mechanical engineer will be designing the system, including ventilation.

6

Concrete masonry walls

Concrete masonry is one of the most familiar building materials used today and one of the most common concrete systems for building walls. Concrete masonry has been used successfully for decades in foundation walls and above-grade walls alike.

Concrete masonry units are generally hollow units that are manufactured with a dry-cast process in a plant. They are typically made of portland cement, aggregate and water. The most common masonry is a gray unit that is 8 inches tall and 16 inches long. However, concrete masonry is available in many different sizes, finishes, and colors.

Concrete masonry walls are widely used for several reasons. Masonry construction tends to be economical. Masonry's long track record and its familiarity make it well understood by contractors, building departments, engineers, architects, and building owners. Therefore there is little difficulty having it accepted for projects. Masonry construction also offers some flexibility. The units can be shifted, adjusted, and cut in the field to change wall dimensions and features. In addition masonry has the general advantages of concrete walls, including strength, durability, and air tightness and thermal mass.

Concrete masonry walls are usually built by skilled masons. The masons stack one course on top of another with mortar. This labor is widely available. Different levels of insulation can be added by various methods.

About 2 billion square feet of structural concrete masonry walls are constructed in the United States each year. The majority of this goes to small buildings, including below- and above-grade walls.

History

Masonry is one of the oldest forms of construction used by man. Some structures built of stone, mud brick, and clay brick thousands of years ago are still standing today. However, it wasn't until the beginning of the nineteenth century that concrete masonry became popular. It first took off in the United States, where the masonry units were molded using lime and moist sand. They were cured with steam. Soon after, English engineers took interest in the masonry and started making solid units using lime, aggregate, and boiling water. However, their weight created problems. In the late 1800s, new molding techniques allowed the production of masonry with hollow cores.

Soon after, machines were developed in both the United States and Europe to form hollow masonry units automatically. The demand for the product and the speed of the machines both grew rapidly.

The industry grew, and in 1918 a group of people involved with masonry formed the National Concrete Masonry Association to help promote concrete masonry products and their acceptance and proper use.

By the end of World War II, around twenty percent of all new houses constructed in the United States had masonry walls. The percentage of small multifamily and nonresidential buildings constructed with masonry was higher.

However, gradually these numbers declined. Wood frame construction became less expensive and took more of the market. When higher insulation requirements became common during the 1970s, there were few readily available, economical methods of insulating masonry walls.

By the early 1990s concrete masonry accounted for only about 3 percent of new home construction. Nearly all of this was in Florida. Masonry maintained a much higher share of other small buildings, but lost some ground to frame there as well.

In the mid 1990s things turned around rapidly. Major hurricanes ravaged large areas in the Southeast, and expert surveys indicated that the masonry walls tended to survive the winds with significantly less damage than frame did. This highlighted the strength and disaster resistance of the system. New options for insulation became available. These, coupled with the inherent thermal mass and air tightness of the walls, offered the building owner potentially greater energy efficiency than with many other wall systems. Currently about 10 percent of the new homes in the U.S. are constructed with masonry walls, as well as a large proportion of the other small buildings.

Market

Masonry units are used widely throughout the nation. They are common everywhere in commercial, institutional, and industrial buildings. Their share of the market is particularly high in the South and parts of the Midwest. Almost every type and class of these nonresidential buildings is constructed with masonry, at least occasionally. This includes everything from the most basic functional structures, such as sheds and barns, to highly architectural showcase buildings.

Masonry is also used heavily for basements where basements are common, which is mostly the northern states. Masonry construction for above-grade walls of residential buildings is still concentrated in the Southeast. In Florida alone, masonry accounts for about 84 percent of all single-family homes. However, the system has spread to neighboring southeastern states as well, and occasionally masonry houses are built in other spots across North America.

Advantages to the owner

Masonry walls have the usual advantages of concrete, including strength, durability, air tightness, and thermal mass. Masonry also has an additional advantage that is not always present with other systems. That is design flexibility. It is relatively easy to make changes during construction by positioning the units differently. Masonry can also be less expensive than other wall systems, though this depends on a number of factors. Masonry also offers a number of distinctive and aesthetically pleasing finishes that are created in the manufacturing process. This can eliminate the time spent applying an additional finish after the wall is erected.

Advantages for the contractor

A large advantage to the contractor is the familiarity of masonry. Since it is has been used for hundreds of years and is thoroughly covered in the major codes, few contractors encounter hesitation from building inspectors, architects, or engineers. The product's flexibility is also an advantage for the contractor. Many changes and unanticipated problems may be accommodated by simply shifting the units.

Components

Concrete masonry units come in many shapes, sizes, finishes, and colors. The most common units are gray. Their dimensions are 8 in. (wide) by 8 in. (high) by 16 in. (long) and 8 in by 8 in. by 8 in. Note that these are the *nominal dimensions* of the units. Their actual dimensions are about ⅜ in. less than stated. This allows a space of ⅜ in. on all sides for the layer of mortar that binds the units.

In Canada, standard concrete masonry sizes are very slightly smaller, enough that they may not be mixed with U.S. units at the job. The nominal dimensions are 200 mm × 200 mm × 400 mm, and the actual dimensions are smaller by 10 mm to allow for joints 10 mm wide.

Concrete masonry may also be manufactured with a wide array of distinctive textures. Table 6-1 lists the more popular ones. They can also be manufactured in almost any color. By mixing in aggregates of different colors, coloring the concrete, and applying one of the custom textures, a limitless array of appearances is possible. Such textured and colored units are often referred to as *architectural block* (Table 6-1).

Concrete masonry manufacture involves machine molding dry, zero-slump concrete into desired shapes. The units are then stored and cured using heat and steam. They are manufactured at over 600 plants across North America

Table 6-1. Popular Architectural Block Finish Textures.

Finish	Description
Smooth face	Plain, as it comes out of the mold
Split face	Irregular, fractured surface resembling rough cut stone
Sandblast	Lightly stippled surface like finely cut stone
Ground face	Ground smooth
Tumbled	Rounded at the edges and corners like old stones
Slump	Bulging surface like adobe brick
Glazed	Coated with a permanent glaze that is usually bright and high gloss

and shipped to the job site. Concrete masonry units are most commonly made of portland cement, graded aggregates, and water. However, mixtures may contain other ingredients such as air entraining agents, coloring pigments, pozzolanic materials, and water repellents. The standard concrete units have aggregates made from sand, gravel, and stones.

The $8 \times 8 \times 16$ unit is the one most commonly used. It is sometimes called a *full block* or *stretcher*. The $8 \times 8 \times 8$ is called a *half block*. Blocks with reduced webs in the middle are called *U blocks* or sometimes *bond beam blocks* or *lintel blocks*. They are installed where the wall requires horizontal reinforcement. There are many other shapes and sizes available for special purposes (Figure 6-1).

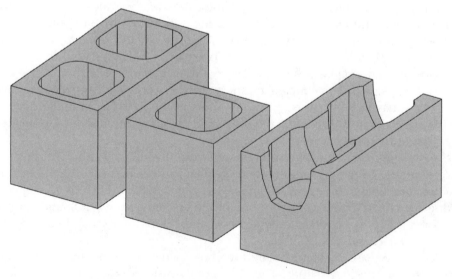

6-1 *Full (left), half (center), and U block (right) masonry units.*

Masonry mortar attaches one unit to another. Mortar is made from one or more cementitious materials, masonry sand, and water. The water content can be varied to adjust the consistency.

Masonry walls are reinforced with conventional steel reinforcing bar and construction grout.

The wall assembly

Masonry walls are almost always laid in a pattern called *running bond*. In this pattern, each *course* (a horizontal row of masonry) is offset from the one below by at least one fourth the length of the unit. In practice they are usually offset by one half the length of the unit. In contrast, with a *stack bond*, the units on each course are offset by less than one-fourth the length of the unit. In fact, they are usually lined up directly over the units below. Between the units on all sides is a ⅜-inch layer of mortar (Figure 6-2).

The units are stacked up to the sides of openings for windows and doors. This may require the use of half blocks on some courses to end at the desired side of the opening.

Masonry walls may be either reinforced or not, depending on the wall strength required. In reinforced walls, parts of the wall are filled with grout,

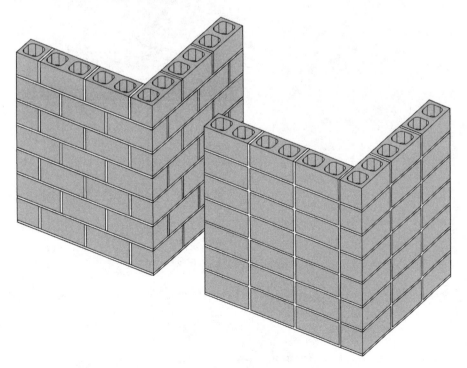

6-2 *Running bond (left) and stack bond (right).*

and the grout embeds reinforcement. It is possible to fill one vertical column of cavities with grout, with a vertical bar embedded in the center. Using lintel units it is also possible to fill one horizontal course, with an embedded horizontal bar.

In a reinforced masonry wall, typically there is one filled cell every 16 inches to four feet along the wall, depending on the level of strength required. There is also one horizontal course filled at the top of the wall or at the top of each story. A reinforced horizontal course is called a *bond beam*. There is also usually a reinforced horizontal member over the top of each opening. This is called a *lintel* (Figure 6-3).

In an unreinforced wall, there may be grout and reinforcement at a few select locations like the corners and around openings. However, there are few if any regular bond beams or reinforced cells in the main body of the wall.

Masonry walls may be either single-wythe or double-wythe. When they are double-wythe, the inner wythe is usually standard concrete masonry units and serves as the structural wall of the building. The outer wythe is most often brick.

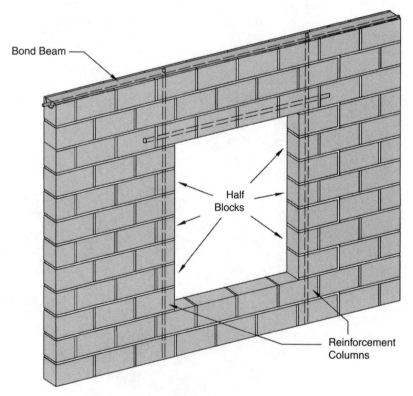

6-3 *Features of a reinforced masonry wall.*

Insulation may be in any of four different locations in the wall. The most common is *interior insulation*. An interior insulated wall has studs or furring strips attached to its inner face. They are typically vertical and every 16 inches on center, as with a conventional frame wall. Between them is insulation. A second insulation arrangement is *cavity insulation*. This consists of insulation inserted into the cavities of the units. The insulation is usually rigid plastic foam. A less-used insulation method is *exterior insulation*. Foam sheet insulation is held in place on the outside of the wall with fasteners. The exterior finish goes over the foam. Double-wythe walls may be insulated with sheet foam placed between the wythes.

The position of electrical cable and boxes varies with the wall type. When the wall is insulated on the interior, the electrical is installed between the studs or furring strips, much the way it is in a frame wall. For other walls, it is possible to run it through the empty cavities of the unit. Typically, rectangles are cut out of the inner face of the unit and the boxes inserted into them. Conduit runs through the cavities to the boxes, with the cable inside the conduit.

Cost

Masonry construction costs vary widely by region because of differences in labor rates, weather conditions, and the cost of making the units themselves. A masonry subcontractor in parts of the South might charge four or five dollars per square foot for an uninsulated masonry wall on a small, low-height building. In the Northeast and Northwest, the charge would be closer to six or seven dollars per square foot. It may be somewhere in between in the Midwest.

Wall cost can be the same or close to frame wall costs in the more economical areas. However, in the Northeast and Northwest it will typically be higher. Table 6–2 includes a typical breakdown of costs for the walls of a small, low-height building, using gray masonry without insulation.

Table 6-2. Approximate Costs of Concrete Masonry Wall Construction.

Item		Cost per Square Foot*
Labor		$2.50
Materials		$3.10
Mortar and grout	$1.00	
Rebar	$0.35	
Masonry units	$1.00	
Total		$5.60

*All costs in U.S. dollars per gross square foot wall area, not including the subcontractor's markup.

The cost of masonry construction rises significantly with the height of the building. Taller walls require more extensive scaffolding and more time in lifting materials into position. However, the scaffolding in multi-story buildings may be set on the floors that are constructed as the building goes up.

Complex wall features and unusual dimensions have a significant effect as well. Masonry construction is most economical with straight walls, right angles, and dimensions in increments of 8 inches. This matches the dimensions of typical units. Irregularities and other dimensions may require cutting large numbers of pieces and assembling the wall by unusual methods. This adds time and cost. Specialty units (such as 45-degree corner units) may be available, and these reduce cutting time.

Masonry labor rates vary widely as well. They are generally lowest in the South, where there is a large supply of masons and the crews can work year round. They are higher in most parts of the North.

Code and regulatory

Concrete masonry walls are thoroughly covered in all major building codes. The International Building Code (IBC) covers concrete masonry in Chapter 21 (entitled "Masonry"). The International Residential Code (IRC) covers it in Sections R606-R609 ("General Masonry Construction," "Unit Masonry," "Multiple Wythe Masonry," and "Grouted Masonry").

Concrete masonry walls for nonresidential construction are very familiar to the local building departments. They are familiar for home construction in many parts of the Southeast as well. Generally no product documentation is required as long as the building falls under the guidelines of the code. Building officials routinely check to see that the required reinforcement is in place before the grout is placed.

In some areas, the use of masonry for houses is rare. When building a house in these areas, it is more common for building inspectors to have questions or ask for custom engineering. Pointing them to the relevant sections of the IRC can help provide them with the information they need.

The key building codes in the United States require that concrete block meet the specifications of American Society for Testing and Materials (ASTM) standard C90. It states requirements for:

- dimensional tolerances
- minimum face shell and web thicknesses
- minimum and maximum water absorption
- maximum linear shrinkage

Meeting these specifications is so generally followed that the sellers and users of masonry generally do not even discuss it. Unless otherwise stated, the product is assumed to be manufactured to this standard.

In Canada, masonry in the National Building Code of Canada (NBC) is covered mainly in Part 4 (Structural Design), Section 4.3.1 (Plain and Reinforced

Masonry), and in Part 9 (Housing and Small Buildings), Section 9.20 (Above-grade Masonry).

Installation

The masonry crew normally consists of about half skilled masons and half laborers. The masons perform precise measurement tasks and set the units in position. The laborers handle more lifting, moving, and mixing. The crew may be as few as three workers on a very small building. For larger projects there are usually at least a dozen, and tasks are divided so that work can proceed most efficiently.

Measurements and planning

Before installing any units, the crew transfers measurements from the building plans to the foundation below, which is usually a footing or floor slab. This begins with marking the locations of the corners. Lines are drawn between the corners. This establishes the line to which the face of the unit will be laid. Locations of openings and control joints are also marked. The masons may set the units of the first course in position without mortar to ensure measurement is correct. The crew may install poles called *storey poles* at each corner to insure that the units there are correctly aligned and stacked to a precise vertical position. These are often marked at the course heights so the mason can stretch a line between them to show the correct top of each course.

Setting the units

The first course is set on the foundation in a bed of mortar. Generally a mortar bed between ¼-inch and ¾-inch is needed to make up for the imperfections in the footing or slab. The foundation is usually fitted with dowels. These dowels are positioned to be in cavities that will be grouted.

   The first course is set precisely level. The mason places mortar on the ends of each unit as well and butts the units together (Figure 6-4).

   The upper course goes over the one below in much the same way. In a running bond, the position of the units is shifted by at least one-quarter block on each course.

   When the wall reaches the height of the bottom of the window opening, the masons begin to set the masonry up to the sides of the opening and continue laying units. When they reach the top of the opening, a lintel goes across the top to bridge from one side to the other. The lintel may consist simply of lintel units that will be filled with grout and reinforcement later to create a structural beam. In this case the opening will likely have a buck or blockout installed to support the masonry temporarily. The lintel may also be formed out of a steel angle with conventional units set on top, or a precast concrete beam set over the opening.

6-4 *Concrete masonry wall under construction.* Portland Cement Association

Before the walls get to eye level, the crew sets scaffolding to walk on and hold materials within easy reach while they work on the upper courses.

After the masons set all the units to the top of the wall, they drop the vertical reinforcement down the cavities with dowels. Horizontal reinforcement goes around the top course to form a bond beam. The reinforcing bars are wired in position.

Grouting the wall

The cavities with reinforcing steel are filled with grout. This may be done when the wall is laid to the top. In this case it is called *high lift grouting*. It may also be done when the wall is only four feet tall or less. This is called *low lift grouting*. Low lift grouting is especially common when there is a bond beam lower in the wall so that it can be easily filled. The grout is usually placed with a small concrete pump that pushes the material through a hose. As it is placed, one worker consolidates it with a vibrator.

Finishing and bracing

The masons typically check for correct alignment by placing a straightedge diagonally across the wall. If the spacing of head joints is correct, all the edges of the unit will be touched by the straightedge. They also check all the mortar joints to ensure that there are no holes or gaps. Any gaps that do appear they fill with fresh mortar.

After the mortar has achieved adequate stiffness, the workers *tool* the joints. This means they press the mortar inward with a rounded blade called a *jointer*. This compresses the surface of the joint, improving the seal and making a stronger bond between mortar and unit. It also produces a concave joint.

Temporary bracing may be installed on stacked walls to prevent high winds from toppling them. Whether it is required depends on the height of the wall and local wind conditions. Most masonry bracing consists of diagonal kickers every few feet along the wall.

Connections to floors and roofs

The connection of other concrete floor and roof systems is discussed in later chapters covering those systems.

Frame floor decks may be attached with the ledger method. The anchor bolts may be embedded in a bond beam. The crew installs the bolts in holes in the face of the masonry before grouting the beam. There are also special bolts that are designed to attach to a hollow unit in an ungrouted area.

An alternative for connecting a frame floor is with the *pocket* method. In this method, notches are cut in the inner face of the units. The floor joists are set directly in those notches. Extra measures are necessary to secure the joists in those pockets.

Frame roofs are usually attached by the top plate or hurricane strap method. The connectors are generally embedded in the grout of the bond beam on top. If the top of the wall is not grouted, there are various metal connectors that connect to the masonry directly or to anchors in the joints.

Interior walls, fixtures, and wallboard may be connected directly to the masonry with concrete nails or screws. If the interior of the wall is insulated, fixtures may also be attached to the furring or studding.

Connections to doors and windows

Doorframes are often attached by way of a buck. The doorframe is nailed or screwed to the buck just as it is to wood framing. Some doors are designed to attach directly to the masonry. In this case no buck would be installed.

There are also windows designed to be nailed to wood. For these a buck is installed. Others are designed to be attached to masonry with self-tapping masonry screws through the frame. In this case there is no buck. In some areas special *masonry windows* are popular. These are sized to fit precisely into a standard masonry opening. Special units called *sash block* are installed along either side of the opening. They leave a vertical groove on the inside edges, and the windows have flanges that slide into the groove.

Architects

Most architects have very little problem designing with concrete masonry. Most have at least a basic familiarity with it.

Some architects with less masonry experience design buildings with odd dimensions. Plans that stick with dimensions that are multiples of 8 inches can be built without cutting units. When other dimensions are used, the cutting required increases construction time, cost, and waste.

Engineers

Engineers are not normally required on houses in areas where masonry homes are common. Engineering is more often required on larger buildings or on houses in areas where those are unfamiliar.

Most engineers are familiar with designing for masonry. The requirements are in many engineering texts and the building codes. The engineer may need to do more work and may charge more for building types that are unusual in the local area.

Training

Apprentices generally train for two to five years before they are considered skilled masons. Formal training is available from trade unions and vocational programs, and this is supplemented by working on a crew. Some masons get their training strictly on the job, starting as a laborer and working their way up.

The skill required is high enough that a successful building project depends on retaining an experienced crew. There are tens of thousands of qualified crews throughout North America.

Maintenance and repair

Architectural block is typically manufactured to be exposed without any regular maintenance. It usually includes an integral water repellent, and it is usually recommended that a similar, compatible repellent be used in the mortar. Post-applied sealants may be recommended as well. Other masonry should be covered with a finish that will protect it from the elements; this can include paint, stucco, etc.

Occasionally a visible crack may form in the wall. This should be looked at by a qualified contractor. If the crack is not on a structural wall, it can usually be simply sealed and refinished. However, if the crack does upset the structural integrity of the building, then the section may have to be removed and replaced. However, this is rare.

On single-wythe walls, if detail work on the walls is not properly performed, water may find its way through the wall. Correcting this involves tracking the leak back to the error in the wall and correcting it. This is rarely a concern with a double-wythe wall because the cavity between wythes will generally drain out any entering moisture.

Energy

The R-value of a concrete masonry wall can vary widely. Uninsulated, a standard masonry wall has an R-2 to R-3. Insulation in the block cavities can raise this to an estimated R-4 to R-6. Furring strips with insulation can raise it to R-5 to R-8. That is normally considered acceptable in very temperate climates of the South. Studding and insulating the interior raises the R-value to an estimated 10 to 15. Exterior insulation and insulation in a double-wythe wall depends on the insulation thickness chosen, but typically also produces a wall with an R-value of about 10 to 15.

Separate from the R-value, masonry walls have high thermal mass and good airtightness. This will raise energy performance above other walls that have the same R-value but are lighter and less tight. In Florida, interior furring with fiberglass is common. It has been found in this location to provide similar energy efficiency to a standard frame wall filled with fiberglass. In more northern climates, it is sensible to include more insulation. The benefits of thermal mass will be lower there.

Aesthetics

The most popular and least expensive finishes for masonry are plain paint or stucco. In areas where stucco is common, it has been raised to something of an art form. Foam moldings may be formed first, and stucco is applied over the foam to create relief on the wall. The stucco may be integrally pigmented or painted to create any color.

The units themselves may be produced with an architectural finish, as described under Components. Architectural block costs more, and the labor to install it properly may be more expensive. However, the results can be striking and different from what is available with any other common siding or finish. It is also usually maintenance free.

On the interior of the wall, furring and wallboard is most common. When the masonry is left exposed, its natural finish is perfectly functional. However, it may also be painted.

Key considerations

Quality masonry work is critical to a successful building. Significant effort often goes into finding only qualified crews and sometimes waiting on their availability, if necessary.

With a single-wythe wall, proper attention to details is important for good moisture control. These details include such things as well-filled mortar joints and proper masonry work and flashing at openings. Depending on the climate, it may also be important to provide weeps at the bottom of the wall to drain moisture that may get into the cavities.

It is also important that exposed masonry be protected to extend its life and add a barrier to moisture. Architectural block is usually manufactured with an integral water repellent. Other masonry is routinely covered with a finish such as stucco or a concrete paint or both.

Availability

Concrete masonry is almost always close by throughout the United States and Canada. It is manufactured in approximately 600 plants that are placed near the areas of use. It is sold through thousands of masonry supply houses and countless other construction products outlets.

There are thousands of masonry crews throughout North America. On occasion, in a local area, there may be more work than the nearby masons are able to perform. At these times, it may be possible to secure crews from farther away, or it may be advisable to wait for the locals to become available.

Support

Concrete masonry is so commonly understood that the local contractors, architects, and engineers generally can handle any questions that arise. For broader issues, the National Concrete Masonry Association (www.ncma.org) provides extensive technical and background materials. It also employs a skilled staff that can help on many problems and refer users to specialists to help with others. There are also many state and local masonry associations that can be helpful. The NCMA maintains lists of these.

In Canada, the responsible trade association is Masonry Canada (www. masonrycanada.ca).

Current projects

At any given time there are thousands of masonry projects underway across North America. They can be found by asking local contractors and masonry suppliers.

7

Insulating concrete form walls

Insulating concrete forms (ICFs) are hollow foam blocks or panels that are fitted together into the shape of the exterior walls of a building. The crew fills the hollow cavity in the center with reinforced concrete. This provides the structure. The foam stays in place to act as insulation and a backer for finishes (Figure 7-1).

Insulating concrete forms have become popular and have grown rapidly because they have most of the advantages of concrete walls, plus a few. Because of the double layer of foam, ICF walls have unusually high R-values, and therefore high-energy efficiency. They are also flexible in the field because they can be easily cut and glued.

Because ICFs have faces of foam, they must generally be covered on both sides. The foam alone is not highly durable.

The Portland Cement Association has forecast that in 2005, over 200 million square feet of the forms will be sold in the United States. In that year, they are expected to be used in about 80,000 single-family houses, over seven percent of new single-family home construction. In addition, about one-third that volume of the forms will go into nonresidential buildings. ICF sales and construction rates in recent years appear to be growing 20 percent to 30 percent per year.

History

Most observers agree that the first insulating concrete forms appeared in Switzerland soon after World War II. Two inventors developed a product shaped like a concrete block that was made of a mixture of treated wood fibers and portland cement. Stacked up by itself, the block was not strong enough to serve as the structure of buildings, so the inventors decided that its hollows would

7-1 *Worker setting insulating concrete form to construct a wall.* American ConForm

be filled with reinforced concrete. The block itself stayed in place and provided insulation, just like ICFs today.

In the 1940s and 1950s, chemical companies developed plastic foams. In the 1960s, it became apparent that something like the Swiss form could be made out of them. In the late 1960s a Canadian inventor came up with a block made of foam, sold it, and licensed it to others to produce as well. Some Europeans developed similar products as well.

During the 1980s and 1990s, a large number of North American companies designed their own variations on the basic product. Some were similar to the original Canadian form. Some were different, more like panels or planks than like blocks.

In the mid-1990s, the Insulating Concrete Form Association (ICFA) organized to represent ICF producers and others interested in the progress of the industry. With the Portland Cement Association, the ICFA began systematic research and promotion programs for the product. These included the engineering work needed to get ICFs into major building codes. Meanwhile, the many new companies that entered the business developed improved products and educated thousands of contractors in correct installation. Total sales grew 50 percent to 100 percent per year for a time, before easing to their current rate.

Table 7-1. Approximate Breakdown of ICF Sales.

Application		Share of Total Sales
Single-family homes		70%
Houses with basement/foundation only built of ICFs	20%	
Houses with most/all exterior walls built of ICFs	50%	
Commercial and multi-family buildings		30%

Market

Although exact figures are not available, sellers of ICFs estimate that total current sales break down about as listed in Table 7-1.

In the early years, the forms were used mostly for basements of single-family homes. Over time a growing percentage of homebuyers had ICFs installed for all their exterior walls, all the way to the roof. This trend appears to be continuing.

Insulating concrete form home sales are greater among higher-end homes. However, they have steadily moved into the middle of the market, as the product has become better known and construction has become more efficient. Developers have begun to use ICFs for construction of mid-size developments of fifty to 150 homes. They market the homes as superior construction and the developments as more desirable.

Commercial and multi-family construction is a growing percentage of total ICF use. In many cases a building in this class is less expensive to construct with ICFs than with alternative wall systems. This is often true for buildings that house people or products that must be kept under precise environmental control. Some popular applications are apartment and condominium buildings, hotels and motels, retail, movie theaters, and wineries and agricultural storage buildings.

Advantages to the owner

ICF walls have most of the advantages of a reinforced concrete wall, including strength, resistance to natural disasters, rot, mold, mildew and insects, and sound attenuation.

ICF walls also have high R-values that provide very high-energy efficiency.

The foam faces are not durable enough to remain exposed. Impact can dent them. Light will cause the foam to oxidize over time. This turns the surface material into a yellow powder that eventually wears away. Therefore the walls generally require a finish. In high-traffic situations or heavy-duty applications it may be advisable to cover the walls with high-durability finishes.

Foam does not provide a food source for animals or insects. However, ground-dwelling termites have burrowed into foam in some situations. Therefore, building codes and good practice require that ICF walls below-grade have some form of termite deterrent when used in high-termite areas.

Advantages to the contractor

For contractors, one advantage of ICFs is flexibility in construction. The foam is relatively easy to cut and glue back together into desired wall shapes. The forms can also be changed relatively inexpensively right up until the time concrete is placed into them.

The forms are also light in weight. Workers may be less tired at the end of the day and experience fewer physical problems. One worker can handle the parts for the form wall. Some contractors find that they can send one person to a site to erect smaller projects such as stem walls.

Insulating concrete forms are also readily learned and used by carpenters. This has allowed general contractors with carpentry crews on payroll to give additional work to their employees, rather than contracting it out.

Components

All insulating concrete forms consist of two parallel layers of foam or similar material held a constant distance apart by crosspieces. Beyond that, there is considerable variety.

The two major categories of ICFs are *blocks* and *planks*. Blocks are hollow, rectangular units that are set into the wall one at a time. A few of these are all one molded piece of the foam material with a shape something like a concrete block. Others have only the two faces made of foam. The crosspieces running between the faces are light-gauge steel or plastic pieces called *ties*. The ties hold the two faces of foam a constant distance apart (Figure 7-2).

Most blocks are 16 inches high and 48 inches long. However, many companies offer larger blocks, and a few sell a smaller unit. The foam faces vary in thickness from about $1\frac{7}{8}$ inches up to $2\frac{3}{4}$ inches. Blocks are also available with different cavity thicknesses so that the contractor can create walls with different concrete thicknesses. The most-used cavity thicknesses are 6 inches and 8 inches. However, others are available.

Plank systems consist of board-shaped pieces of foam referred to as *planks*. They have special notches along their edges. Fitting into these notches are crosspieces made of plastic or steel. These are either ties like the blocks have, or long truss-like units called *rails* that serve the same purpose (Figure 7-3).

The foam of most ICFs is expanded polystyrene (EPS). However, a few are other materials. Some blocks use a mixture of portland cement and foam beads,

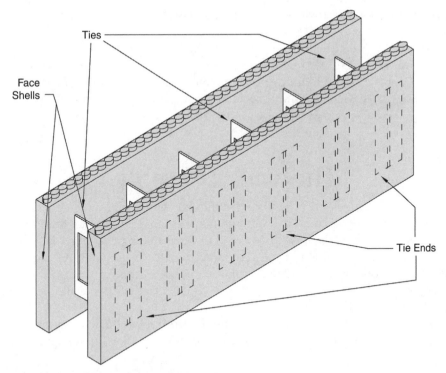

7-2 *The parts of a typical ICF block.*

7-3 *Worker installing the parts of a plank system.* Quad-Lock Building Systems, Ltd.

or portland cement and wood fibers. Some plank systems offer extruded polystyrene (XPS).

Also in the wall are conventional construction concrete and conventional steel reinforcing bar. Also widely used with ICFs is *low-expanding foam adhesive*. This is a liquid foam dispensed from a gun. It expands slightly when dispensed. It is used to glue form units or pieces of foam together where that is necessary. It may also be used to fill in gaps or gouges that might occur.

The wall assembly

An insulating concrete form wall sandwiches a layer of reinforced concrete between two layers of foam. The amount and positioning of the reinforcing bars depends on the loads that will be placed on the building and other factors related to the engineering and the required strength of the wall. As a general rule, there is one line of horizontal reinforcing bar every few feet up the wall, a line of vertical rebar every one foot to four feet along the wall, and additional bars around each opening.

Door and window openings are lined with bucks. Bucks may be made of 2× lumber, or of plastic extrusions that were developed specifically for ICFs.

Most of the ICF systems have the steel or plastic ties that can receive fasteners such as screws. Depending on the brand, ties are spaced every 6 inches, 8 inches, or 12 inches on center.

Cost

In small buildings, ICF walls usually cost a little more than standard wood frame walls. In medium and larger buildings where the requirements are frequently higher, ICF walls may be less expensive than the alternatives. For a small building, subcontractors usually charge about $7.00 to $9.00 per square foot. The costs of construction for a typical above-grade project are approximately as shown in Table 7-2.

Labor rates vary with the complexity of the walls. Foundations with a minimum of corners and openings might have a labor cost 20 percent to 40 percent below "average" rates. Walls with many corners, many openings, irregular features, and changes in wall height might have labor rates nearly double those listed here. Tall walls such as those on some commercial projects that involve extensive scaffolding also have higher labor costs. Differences in wages across regions also affect labor cost.

Form costs are often higher when the wall is more complex. Complex walls generally require more specialty forms, like corner and brick ledge units. These are sometimes more expensive per square foot than straight forms.

Table 7-2. Approximate Costs of ICF Wall Construction.

Item		Cost per Square Foot*
Labor		$1.75
Materials		$5.35
Forms	$3.50	
Concrete	$1.25	
Rebar	$0.35	
Miscellaneous	$0.25	
Pump rental		$0.35
Total		$7.45

*All costs in U.S. dollars per gross square foot wall area.

Wall thickness directly determines the amount of concrete in a square foot of wall. Concrete cost therefore is greater for a thicker wall. Rebar costs are higher in walls with high-required strengths as well.

Code and regulatory

In the United States, construction with insulating concrete forms is covered in the International Residential Code. Section R404 treats foundation and below-grade walls. Section R611 covers above-grade walls.

The International Building Code does not yet include ICFs. Larger and commercial buildings using ICFs are almost always designed structurally by licensed engineers. Most of the major ICF manufacturers have received evaluation reports for their products. Nearly all of these cover their use in small residential applications, but many now cover a wide range of larger and commercial applications as well.

Canada's National Building Code does not yet include ICFs. However, the new edition of the code to be issued in late 2005 will cover them for use in up to two-story buildings under Part 9 (Residential and Small Buildings).

Building officials are beginning to accept ICF construction without requiring special supporting materials. For single-family homes they often require no custom engineering, relying instead on the provisions of the International Residential Code (IRC) and evaluation reports from the International Code Council (ICC) and the Canadian Construction Materials Centre (CCMC). For larger buildings, they are likely to require the same level of engineering they would on similar buildings constructed with traditional materials. However, in some areas that have had little ICF construction, the building officials may still put higher requirements on ICF building projects.

The foams used to make ICFs are held to the same fire resistance standards of other construction foams. The manufacturers are responsible for using only foams that meet these standards.

Installation

Insulating concrete forms are typically installed over a footing or slab foundation. The crew marks the location of the walls on the foundation. They build their bucks in advance of setting the forms. The typical crew size is three to four workers for a small building and six to ten for a mid-size one.

In most small building construction, the construction sequence is similar to that of platform framing. The bottom story walls are completed, followed by installation of the floor or roof, then come the second story walls, the next floor or the roof, and so on.

Form wall assembly

Blocks and horizontal plank forms are set one course at a time. There is no single order to the installation that is best for all projects and forms. However, most contractors set the corners of the first course first and then work toward the middle on each wall to complete the first course. The upper courses are set the same way, one at a time. However, the joints of different courses are staggered much the way they are in the running bond of a concrete block wall.

Somewhere during construction of the first two courses, the crew checks the forms for precise level. An uneven foundation below may throw the formwork off. If this is the case, the crew makes adjustments on the first two courses to correct.

When the wall reaches a height that calls for a line of horizontal rebar, workers lay the bars across the tops of the ties and snap or wire them in place. When the wall reaches the sill height of a window, the crew sets the buck onto the forms. On the higher courses, the crew cuts forms to fit up to this buck.

Once the wall is a few feet high, the crew installs a *bracing-scaffolding system*. This is a special system designed specifically for use with ICFs. It elevates the crew for setting the top courses of forms. It also helps to hold the wall precisely plumb during construction (Figure 7-4).

The crew also fixes the forms in position on the foundation. They may be held on line with blocking on the foundation, by gluing them with foam adhesive, or other means.

After the top course of forms is in place, the crew drops vertical rebar down the wall in the required locations and fixes it in position with wire or other means. The bracing is adjusted to plumb the wall precisely.

Concrete placement

A wide range of equipment may be used to place concrete. Using equipment that allows the operators to control the flow precisely is important because the foam forms do not have strength as great as conventional wall forms. If concrete flows too rapidly or into the wrong place at the wrong time, the forms

7-4 *Bracing-scaffolding in position.* Insulating Concrete Form Association

may be thrown out of alignment. Partly for this reason, many contractors prefer to use a concrete pump even when the work is below grade and they might reach it with the chute from a ready-mix truck. The larger pumps need to have fittings to slow the flow of concrete and smooth it out (Figure 7-5).

The placement proceeds to make work steady without putting undue stress on the forms at any one point. Normally the crew places the concrete in a series of *lifts*. A lift is one pass around the wall that fills it up only part way, usually about three or four feet. A worker with a vibrator trails the person placing the concrete to consolidate it.

Between each lift the crew checks the walls for plumb and adjusts the bracing as necessary to keep them there.

Occasionally a section of foam breaks open. This is referred to as a *blowout*. Blowouts are increasingly rare because forms have become more consistently strong and crews are more experienced. But when they occur, they are easy to fix. Workers on the ground replace the broken section of foam and cover it with a temporary brace. This usually takes less than ten minutes.

7-5 *Placing concrete with a boom pump.* ICF Accessories

Connections

Methods of connecting concrete floors and roofs to ICF walls are covered in the chapters on those floors and roofs. Here are the methods of connecting frame and heavy steel floors and roofs.

Floors

Frame floors are usually attached with a ledger. However, because the ledger must rest against concrete and not foam, the usual ledger method must be modified. One popular option is to cut out squares in the foam behind where the ledger will go. The ledger is outfitted with anchor bolts and screwed into position over the holes before concrete is placed. The concrete flows through the holes to back up the ledger (Figure 7-6).

The other popular option uses a special connector to attach the ledger after the concrete is cast. The connector is a plate of steel with a right-angle bend in it. This is a special connector designed for ICF construction. It is inserted through a slot in the foam so the concrete embeds the long leg of the connector. Workers later attach the ledger with special self-drilling screws to the short leg, which lies flush against the foam face (Figure 7-7).

In larger and commercial buildings, floors are increasingly attached by embedding steel weld plates or bolt plates in the concrete. Steel floor components are then welded or bolted to the plates after the concrete has hardened.

7-6 *Ledger in position before concrete is placed.*
Quad-Lock Building Systems, Ltd.

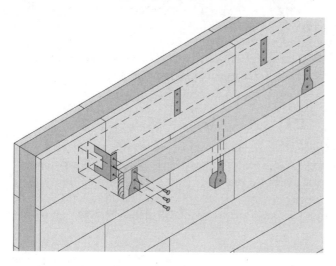

7-7 *Connection of a ledger to a wall with right-angle connectors.*

Roof

Light frame roofs are almost always attached with a top plate or hurricane straps, as on other concrete walls.

As with floors in commercial and larger buildings, steel roofs are coming to be attached with weld plates or bolt plates embedded in the walls.

Other connections

Interior wallboard is generally attached by screwing it to the tie ends. Fixtures may also be attached to the ties. If they fall between ties, there are light-gauge steel products that can be installed in the wall before the wallboard goes up to act as a fastening surface. This is much like the use of blocking to attach fixtures between the studs in a frame wall.

For heavy fixtures, it may be necessary to go through the foam with concrete connectors.

Architects

Most architects adapt easily to designing with ICFs. The product is relatively flexible, so there are few special design rules to learn. Insulating concrete forms can be cut to achieve almost any dimensions relatively economically. Irregular features, such as non-ninety-degree corners, curved walls, and non-rectangular openings are also relatively easy to create.

Architects, as a group, are not as familiar with ICFs as they are with some other materials. However, ICFs are well documented. Suppliers provide detailed manuals. Some publish books of plans designed for their product. Introductory publications are available from the Insulating Concrete Form Association (www.forms.org) and the Portland Cement Association (www.concretehomes.org). They can acquaint designers with most of the product features they need to do their work.

A growing number of architects now have experience working with ICFs or basic knowledge of them. The independent web site www.icfweb.com includes a listing of architects in its directory of professionals.

Engineers

When structural engineering is required, engineers generally work well with ICFs. ICFs create a reinforced concrete wall, which is something most engineers are thoroughly trained in. They may not be accustomed to using it in very small buildings, but the provisions of the International Residential Code cover this. Engineer-to-engineer support is available from the major forms suppliers.

Training

Experience shows that construction proceeds best when the head of the ICF crew goes through a formal training program and the crew gets assistance from an experienced person on its first project.

Most major ICF suppliers offer classes to train users in proper installation of their product. These last one or two days and carry a tuition fee of $100 to $300. The United Brotherhood of Carpenters (www.carpenters.org) has thorough courses on ICF construction, which are available to UBC members at many of the locals. The Cement Association of Canada (www.cement.ca) has developed curriculum for ICF installers that it provides to trade schools and other training organizations. Contact the CAC at (613) 236-9471 for details of where it may be offered.

Like many construction tasks, installers cannot become fully skilled in ICF construction until they actually do it. Having an experienced ICF installer assist on a crew's first job is common. Often the local distributor who sold the forms can supply or recommend a good person.

Maintenance and repair

Insulating concrete form walls require little or no routine maintenance. They are supposed to be covered by finishes that protect them from the elements. If finishes detach or wear off in spots, they should be replaced quickly.

There are some reports of damage to the foam at ground level. This occurs most often at the building's corners. It is possible when the wall is not well covered by a high-impact finish material. Impact from such things as lawnmowers or foot traffic can break through a weak finish and dent or gouge the foam. The repair is simply to pull away any loose pieces, fill in the gap, and replace the finish.

Energy

There is general agreement that using ICF walls instead of typical frame walls will reduce a building's energy consumption. Estimates of this vary widely, but most suggest the savings are 20 percent or more.

The R-value of an ICF wall is typically about 20. The air tightness of the house is also high. Measurements suggest that if a house has ICF walls and typical windows, doors, and roof, the rate at which air flows into it from the outdoors will be 10 percent to 30 percent lower than with typical frame walls.

Like all concrete walls, ICF walls have a relatively high thermal mass. Some people have argued that the effect of the thermal mass on indoor air temperature will be reduced because it is isolated from the air with a layer of insulation. However, engineering simulations suggest that there will still be an energy-saving effect, even though it will be somewhat lower than it would have been if the concrete were exposed.

Sustainability

The major environmental benefit cited for ICF construction is its energy efficiency. Producing the foam requires oil, and significant energy is consumed in production of the cement used in the concrete. However, comparisons have indicated that the energy saved in the building's fuel bills outweigh the initial extra energy within a few years.

Like many concrete products, ICFs are also frequently cited as a method of reducing the use of trees for lumber.

Using ICFs frequently enables a building to qualify for LEED points for energy efficiency. It may also help to qualify for a credit for use of local materials. The concrete almost always comes from within the 500-mile radius allowed by the LEED system. Usually the forms do as well.

Using ICFs may or may not help a building qualify for a credit for recycled content. The concrete may contain some recycled material, such as fly ash. Rebar sometimes is made with recycled steel. It would be necessary to consult the manufacturers to determine either of these things. Insulating concrete forms are rarely made with recycled plastics, although a few are.

Aesthetics

Nearly all popular finishes can be installed onto ICF walls at about the same cost that they can be installed onto wood frame. By far, the most popular finish material for the inside is conventional gypsum wallboard.

Finishes applied with a trowel, such as stucco, are usually a little less expensive to install over an ICF than over wood. Over wood there is the additional step of first putting up a weather-resistant membrane or layer.

Any exterior finish material attached with fasteners to wood frame, such as vinyl or wood sidings, also attach to ICFs with screws into the ties. The cost and time involved are about the same.

Masonry is installed with brick ties screwed to the ICF ties, just as the brick ties are attached to wood frame.

Key considerations

The most common serious errors resulting from poor ICF construction work are out-of-plumb walls and concrete voids.

Out-of-plumb walls are prevented with a quality bracing system that is properly adjusted. Experienced ICF contractors recommend a modern metal bracing system with continuous threaded adjustment. It must be adjusted at several points during the concrete placement.

Research shows that proper consolidation with a vibrator eliminates nearly all significant voids. It is important that the contractors perform this consolidation.

As with some other wall systems, with ICFs, HVAC contractors tend to oversize their equipment. The usual solutions apply: shop around for a qualified HVAC contractor or, in the case of larger buildings, count on the mechanical engineer to estimate the building's needs more accurately.

If fire burns through the finishes on an ICF wall, it can cause the foam to give off fumes similar to the smoke of burning wood. The concrete structure may survive fire better than a wooden wall. However, because of the hazards of smoke, it is important to cover ICF walls adequately. Most codes require that foam be covered at least in all inhabited areas of the building. Bear in mind that all fire precautions recommended in a conventional building need to be heeded in an ICF building as well.

Availability

Insulating concrete forms are sold throughout the United States and Canada. In most major market areas, multiple brands are available. In a few rural areas there may be no local distributors. However, manufacturers will generally ship forms to job sites. Manufacturers and distributors are listed on the web site of the Insulating Concrete Form Association (www.forms.org) and the independent web site www.icfweb.com.

Crew availability

Most areas of the country now have experienced ICF crews willing to work on a subcontract basis. To find them, ask local ICF distributors or look at the listings of contractors on www.forms.org or www.icfweb.com.

Assembling a new crew is also feasible. Ideally, the crew will have at least one skilled carpenter and at least one person experienced with concrete. These two could be the same person. The lead person on the crew should also be trained in ICF construction.

Many ICF crews are simply crews from other trades that hire a new person or do some training to cover the key skills they lack. Insulating concrete form crews that build homes are frequently former carpentry crews. In commercial work, they are more often former commercial concrete crews.

Support

Most ICF manufacturers have a network of local distributors. These serve as the first source for technical support for contractors. They are usually readily available for answering questions and visiting job sites.

If no local distributor is available, support is usually available by calling the company's headquarters. Most major ICF manufacturers have engineers and experienced contractors on staff to answer questions.

For certain broad questions it may be helpful to call the related trade associations, including the Insulating Concrete Form Association (www.forms.org), the Portland Cement Association (www.concretehomes.com) in the United States and the Cement Association of Canada (www.cement.ca).

Current projects

It may be possible to find ICF projects that you can view by asking local forms distributors or ICF contractors. These are listed on the site of the Insulating Concrete Form Association (www.forms.org) and the independent site www.icfweb.com.

8

Precast concrete walls

Precast walls are panels of concrete cast in a factory. They are shipped to the job site and hoisted into position to become the exterior walls of a building. The panels are connected to the foundation, floors, and roof to create the total building structure. They can be produced with varying amounts of insulation for different levels of energy efficiency (Figure 8-1).

Precast panels have been popular in large and mid-size buildings for decades. Use of precast panels is now growing in small buildings because of their potential advantages. In addition to the usual advantages of concrete walls, they can be erected in a very short time. Fabrication in the factory can provide a high level of quality control and reliability. The amount of insulation can often be adjusted to produce a basic, economical wall or an especially energy-efficient one. There are different panel configurations available to meet different requirements. A variety of finishes may be achieved at the plant, and some other finishes may be added later.

For economical small precast buildings it is important to have a producer in or near the region of the project to minimize shipping. Wall panels need to be ordered in advance. It is also important to minimize changes in the design after the panels are produced.

The new precast wall systems for small buildings were used to construct the above-grade walls of an estimated 1,000 to 2,000 single-family homes in the U.S. in 2004. About half as many other small buildings were estimated to be constructed with them.

History

Precast objects were made for various small items in the late 1800s. As methods and equipment were developed for handling larger and heavier objects, factories began to make and sell those as well.

8-1 *Crew setting precast wall panel in position.* Superior Walls

In North America, entire wall sections made in a plant and shipped to the site appeared in the 1900s. In 1944 companies and groups involved in precasting concrete met as the "precast committee" to discuss common interests. As precasting became more widespread, these groups organized formal trade associations. In 1956, precasters and related companies organized the Precast/Prestressed Concrete Institute (PCI). In 1965, similar interests created the National Precast Concrete Association (NPCA).

By the 1960s precast had become a major form of construction. However, most of the wall panels were for large buildings. These were primarily *architectural* panels, which sheathed the exterior of a large building and gave an attractive appearance to the building. They played less of a structural role. They

8-2 *Precasting walls in the factory.* Dukane Precast, Inc.

were usually attached to a heavy steel frame structure. Some small buildings were also constructed with precast panels, but in the 1970s the number was not large (Figure 8-2).

A surge in the use of precast panels for the exterior walls of small buildings began in the 1980s. Systems for small building production came from two different directions. First, inventors in the Northeast created panels for the construction of basements. Trucked to the job site, they were hoisted into place to create all the walls for a foundation in just a few hours. In the early 1990s in Florida, precasters began to design and produce similar panels not for the basement market, but for the above-grade walls of houses and small commercial buildings. The idea spread to other regions.

Meanwhile, some precasters specializing in panels for large buildings created new versions of their product for small buildings. The result has been a variety of different types of precast panels available to fit different needs.

Market

According to current estimates, about 70 percent of the small buildings with precast above-grade walls are single-family homes and multi-family buildings. The bulk of the rest are commercial buildings.

The single-family homes are concentrated in the southeastern United States. There, several manufacturers are each believed to supply panels for over 300 houses per year. The multi-family buildings are spread throughout North America, but are often located in the developed parts of large cities. In these situations, the short time needed to construct the building with precast and the limited amount of space required for construction are important advantages.

The single-family homes include both high-end custom houses and production homes that sell to the middle of the market. For the custom homes the panels are produced to suit the buyer or designer, at a premium price. The production homes follow more standardized plans. This provides some repeatability for the producer, which lowers costs.

In high-wind areas, like Florida, the cost of the homes is often competitive with other common construction methods. Therefore, in these areas the houses carry the appeal of a high-quality concrete structure at a conventional price. This is believed to drive sales in these areas.

The commercial buildings include a wide variety. However, nearly all of these are occupied buildings, where efficient space conditioning and noise control are important. Common uses are in movie theaters, offices, retail stores, and hotels and motels. The precast commercial buildings are widely distributed throughout the United States and Canada.

Advantages to the owner

Precast walls have virtually all the advantages that come with a reinforced concrete wall: strength; resistance to disasters, fire, mold, insects, and vermin; sound reduction; thermal mass and air tightness for energy efficiency; and durability.

Precast walls also have some distinctive properties. When they are designed and fabricated in advance, construction time at the job site can be very short. This is important to owners of such buildings as schools and retail stores, for whom completion times are critical. Wall production in the plant allows for a high level of quality control. The exposed concrete face of precast walls makes it possible to install distinctive and durable concrete finishes that are often desirable.

Many precast wall systems allow for either low, economical amounts of insulation or high insulation levels for an energy-efficient building.

In most precast buildings, it will be important for the owner to commit to the design and resist making changes late in construction.

Advantages for the contractor

There are also advantages for the contractor. The shorter time required on-site reduces risk of delays from such factors as weather and labor availability.

Quality risks are generally also small. The plant certifies the correctness of the panels and is responsible for fixing errors in them. With most of the work done in the factory, it is easy to control the small number of tasks performed in the field. Most manufacturers of panels for single-family homes even send out their own crews to install them, and take responsibility for the correctness of the installation.

For most general contractors, there is little learning necessary to adopt precast construction. For homes, the details of constructing and installing the panels are handled by the vendor. For multi-family and nonresidential buildings, the number of steps and the complexity of the site work are still lower than for many other forms of construction. For the same reasons, the general contractor needs to buy few new tools and make little investment to begin building with precast.

Use of precast panels generates almost no waste on the job site. The manufacturer ships exactly the panels needed, with little or nothing discarded.

When constructing precast buildings, it is important to plan the details carefully in advance. Making changes becomes costly and time-consuming later in the process. Depending on the type of precast system used, it may be necessary to determine the overall configuration of each wall panel. It may also be necessary to specify details such as the location of each electrical outlet.

It is also important to order the panels with sufficient lead time. Depending on current conditions, the precasters may have plant time to produce panels a few weeks after they are ordered, or not until a few months after.

Components

Precast panels are generally rectangular, flat wall sections. They include one layer of reinforced concrete. Most also include a layer of foam or other insulating material.

The panels most often used in small buildings fall into two categories, *hollow core* and *sandwich*. Hollow core panels have tubular hollows in the center of the panel. These reduce weight and material while still providing a strong wall. In the concrete is steel reinforcement. The reinforcement may be conventional rebar or prestressed reinforcing strands. Many hollow core panels are uninsulated. They can be used for unconditioned buildings, or insulation can be added to the interior face of the wall later. Other hollow core panels contain insulation layers built in much like sandwich panels (Figure 8-3).

Sandwich panels themselves divide into two groups: the *full sandwich* and the *half sandwich*. Full sandwich panels have a layer of foam insulation in the middle and a layer of concrete on each face. Generally speaking, the layer of

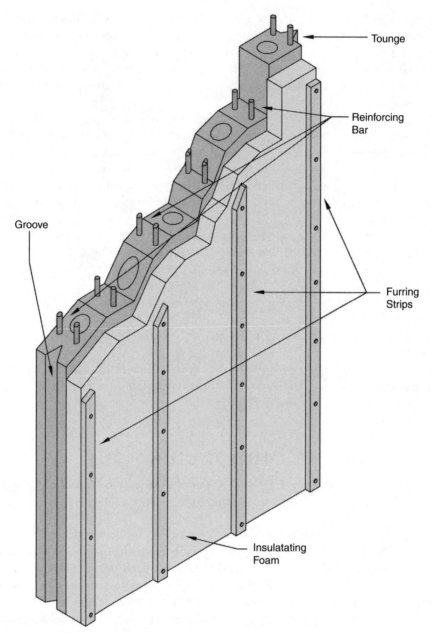

Tounge

Reinforcing
Bar

Groove

Furring
Strips

Insulatating
Foam

8-3 *Hollowcore wall with insulation and furring.*

concrete on the inside face is reinforced with rebar and is structural. It is usually 5 inches to 8 inches thick. It is responsible for bearing most of the loads
of the building. The outside concrete layer serves as the exterior finish and
protects the foam. It may be only two inches thick. Connecting the two con-

crete layers are steel or composite plastic struts or pins. These extend through the foam.

Half sandwich panels have a layer of concrete on the outside only. This is reinforced and structural. On the inside face is the foam layer. Most half-sandwich panels also have wood or light-gauge steel "studs" running vertically up the face of the foam.

Some half-sandwich wall panels include studs, and a layer of foam that does not extend all the way to the interior wall surface. Therefore the foam is recessed while the studs protrude inward from it. This leaves an open bay between the studs, much like a conventional frame wall. That space is available for running electrical cable freely after the walls are erected. The bays may also be filled with further insulation if desired. These features are visible on the half sandwich wall in Figure 8-1.

The foam used is usually either expanded polystyrene (EPS) or extruded polystyrene (XPS). In some full sandwich systems it is polyurethane. With many full sandwich systems the panels can be made with different thicknesses of insulation. The buyer may choose the foam thickness needed to attain the wall R-value desired.

The wall assembly

For small buildings, precast wall panels are fixed in position with either grouted cells or steel bolts.

When grouted cells are used, the edges and tops of the panels are concave. Two panels set next to one another have a hollow column between them. Grout fills this column. Grout also fills the hollow channel running along the top of the wall around the perimeter of the building. This creates a *bond beam* along the wall. Rebar is embedded in the grout columns and bond beam. Typically one bar in each column extends into the foundation below and up into the bond beam on top. Additional bars run through the bond beam around the perimeter (Figure 8-4).

Bolted panels instead have steel plates along their bottom and side edges. Bolts through the plates along the bottom of a panel extend into the foundation and secure the panel there. The plates along the side of one panel line up with matching plates along the side of the adjacent panel. Bolts through these plates secure each panel to the ones next to it.

There are three possible locations for electrical lines. In hollow core or full sandwich walls, conduit and recesses for electrical boxes are usually cast into the concrete at planned locations. In half-sandwich panels, the electrician typically cuts chases in the foam for the cables. In half-sandwich panels with foam that is recessed, it is possible to run the cable in the open bays as in frame walls. The bays may then be filled with further insulation if that is desirable.

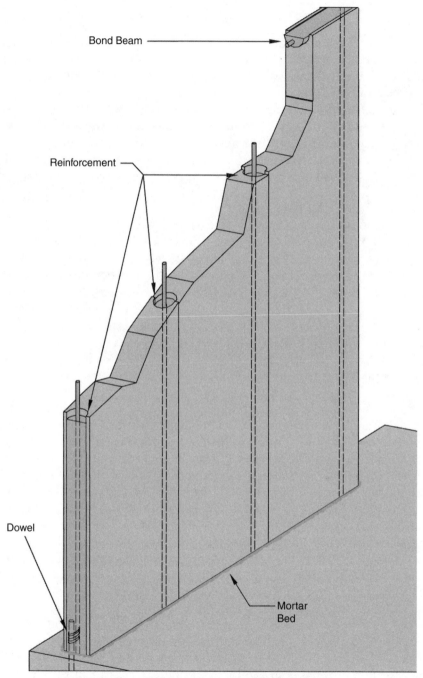

8-4 *Cutaway view of wall made with grouted cells.*

Cost

A typical charge to a general contractor for precast wall panels including installation is $6.50 to $8.50 per square foot of gross wall area. For single-family homes, the panel supplier may provide the installation as part of a package with the panels.

With multi-family and nonresidential buildings, the precaster usually sells the panels to the general contractor without installation. A separate crew installs the panels on-site. This subcontractor charges the general contractor for the installation.

In a small building the cost may be more than a wood frame wall. However, in areas with stringent wind codes, added measures may be necessary on wood frame and increase the cost of frame construction. In these areas precast walls may be about the same price. In larger projects where greater strengths are necessary, the cost of precast walls may be about the same as those of wood.

Table 8-1 includes a cost breakdown for a typical set of bolted, insulated precast wall panels.

Panel and assembly costs vary with the type of wall and the complexity and size of the project. Panels are less expensive if they do not include insulation or finishes. They are also less in large projects that include many of the same design of panel. This repetition provides economies of scale in the panel manufacturing, making the panel lower cost.

Total panel cost is also less when the floor plan is simple. Corner panels and special panels are usually priced higher.

Code and regulatory

In single-family residential projects, the panel manufacturer usually provides engineered, stamped plans for each house they supply. The manufacturer's engineers are also available to answer any questions that the building department might have. With this custom engineering on each building, other support materials are usually unnecessary.

Table 8-1. Approximate Costs of Bolted Precast Wall Construction.

Item		Cost per Square Foot*
Labor		$2.00
Materials		$5.00
Panels	$4.75	
Miscellaneous	$0.25	
Crane rental		$0.35
Total		$7.35

*All costs in U.S. dollars per gross square foot wall area.

On multifamily and nonresidential buildings, custom engineering by the building designers is standard practice, anyway. The panel producer provides engineering data and specifications on their panels. The building designers incorporate these into their analysis and provide the final design and engineering plans.

Precast wall panels are not named in the International Residential Code. In the International Building Code, the general concrete provisions cover the panels. These are based on ACI 318, "Building Code Requirements for Reinforced Concrete" from the American Concrete Institute.

Many producers of precast wall panels for small buildings also have evaluation reports for their systems. These may be helpful for informing building officials about the systems when custom engineering is not available or the building department does not feel that it is sufficient.

The foam used as insulation must meet the general fire resistance requirements of foam used in construction.

Installation

Since precast wall panels are fabricated in a plant, installation on the job site involves few steps and is relatively fast. For an average house, a crew of three or four (not including the crane operator) usually erects all the panels in half a day.

For a commercial building, about four to six workers (not including the crane operator) are usually involved, depending on the size and type of project.

The panel manufacturers typically provide shop drawings showing the correct location for each panel. The crew may put lines on the foundation along the correct edge of the exterior walls. Normally there is a planned sequence to the installation of the panels. It usually begins with the panels at one corner of the building.

Bolted panels

The crane sets the first panel in position. They then install a diagonal brace from the inside surface of the wall to the inside ground or floor. This is adjusted to hold the panel plumb and on its intended line. A worker typically drills into the foundation through the bolt plates along the bottom edge of the panel and then fixes a bolt into the hole. A nut over the exposed end of the bolt tightens it down to the foundation. The next panel along the wall is set in the same way. Then additional bolts are run through the plates on the panels' edges to attach them together. The crew typically works its way around the building in this fashion, one panel after another. When the last panel is up and its edges bolted to either side, the braces can be removed.

Depending on the design and circumstances, the panels may not be set in a simple order of going strictly around the building one panel at a time. Some panels may be set in one area, followed by other panels at the opposite end of the building. This is especially true of larger projects. However, measurements must be carefully taken to insure that all panels will fit precisely when they are finally installed next to one another.

Some installations may call for steel shims to go under panels to compensate for uneven foundations. Typically, one crewmember checks each panel for level. Then the braces are adjusted and shimmed as needed. After the panels are properly set, the crew pushes grout under the edges of the panels to provide continuous, level bearing.

The crew also installs sealant in the joints between panels to close them to air and moisture.

Grouted panels

Grouted panels are set in the same fashion as bolted panels, with a few exceptions. There are usually steel rebar installed in the foundation at the intended locations of the columns. Recall that these are the points where two panels meet side-by-side. The rebar extend up vertically into the space created by the two concave panel edges. Instead of being bolted together, the panels are simply held in position with braces for the time being.

The crew installs a vertical rebar in each column so that its bottom end overlaps the dowel extending out of the foundation. Its top end extends into the bond beam. These bars are wired in place so that they stay securely in the center of the column. A separate line of bars goes around the perimeter of the bond beam.

With all panels in position, grout is placed into the columns and bond beam. After the grout hardens, the crew removes the braces and construction continues.

Multi-story buildings

In a building with multiple stories, the panels for the first story are completely set first. After they are secure, the floor deck may be installed next, and then the upper-story walls. Alternatively, the walls may be installed and secured first, and the floor inserted later.

With bolted panels, the lower-story and upper-story panels are bolted together. Typically, there are steel plates embedded in the top edge of the lower panels and in the bottom edge of the upper panels. These line up so that bolts can be inserted through them and tightened.

With grouted panels, the vertical rebar are sized to extend about two feet above the top of the lower walls. The upper panels are set on top of the lower. Dimensions are set so that the rebar protruding out of the lower walls extend into the columns between the upper panels. The upper panels have concavities along their side and top edges that are outfitted with rebar and grouted, just like the lower panels. The grout encases the rebar protruding from the lower level as well. This connects the two stories structurally.

Connections

Methods of connecting concrete floors and roofs to precast walls are covered in the chapters on those floors and roofs. Here are the methods of connecting frame and heavy steel floors and roofs.

Floors

A common method of attaching floors to the walls is with a so-called "mezzanine" connection. For this connection, steel weld plates are embedded into the panels when they are produced. These plates are just below the intended height of the floor. After the walls are in place, a right angle steel section is welded to the plates. This creates a shelf along the wall. The floor is set on this shelf.

Roof

Light frame roofs are attached by the conventional methods used for concrete walls. For bolted walls, there are typically straps embedded along the top of the wall that are used to secure a top plate. In grouted panels the crew installs connectors in the bond beam for attaching a top plate later. Alternatively, hurricane straps are sometimes used to connect to the roof trusses or rafters directly.

Heavier steel roofing trusses are typically welded to the precast wall. The weld plates are usually cast into the concrete panels at precisely measured locations.

Other connections

If the panels have an interior surface of concrete, interior walls may be attached to the panels with common concrete connectors, such as concrete nails or concrete screws. If the panels used have studs, interior walls can be nailed or screwed directly to the studs. If the interior walls fall between studs, they can fasten to sheet metal strips or plates that bridge two studs.

Lightweight fixtures can be attached in the same ways as interior walls. Heavy fixtures may require concrete connectors or attachment to multiple studs.

Architects

Design with precast walls is familiar to many architects. It is also not difficult to learn or understand. Most manufacturers of precast panels have extensive manuals and documentation that the architect can refer to.

In addition, an architect designing a building will need to work with the precaster to determine the details of the building in advance of construction.

The precaster can alert the architect to any aspects of the design that might be expensive or impractical.

Engineers

On most precast buildings, the engineering of the panels is done by the company selling them. They receive the overall building design and determine the aspects of the panels that are critical to the structure. These include such things as the strength of the concrete, the thickness of the concrete, and the size and placement of the rebar.

The manufacturer may also determine how the panels will be connected. This is depicted on plans provided to the building department and the installers. The manufacturer's design is supposed to create a building structure that meets the building code.

The building owner may instead hire an independent engineer to perform the design of the building. In this case, there is plenty of information for the engineer to draw from. Precast is a form of construction with long use and well-established rules of engineering. Standard engineering software packages include precast modules. The precast manufacturers have complete engineering design guides and some of them have free engineering software available through their web sites. They also have staff to assist.

Training

Often the precast panel manufacturer provides the crew to install its panels. In this case the building's general contractor need not be concerned with assembling and training workers to do the job.

When the GC does take responsibility for setting the panels, usually one or two skilled people on the crew is enough for a small residential or commercial project. The skilled workers serve as supervisors. However, all workers must have safety training.

Many manufacturers of precast wall systems run training classes for installation crews. They may take registration through their web sites. One day of training is usually enough for an experienced carpenter or concrete worker. Local precast associations also sometimes offer training seminars in precast construction. For more information on these opportunities, contact the national trade associations, the Precast/prestressed Concrete Institute (www.pci.org) and the National Precast Concrete Association (www.precast.org).

Maintenance and repair

Precast wall panels are highly durable. The exterior surface is concrete. The interior surface may be concrete or insulation and studs covered with gypsum wallboard. Repairs are rarely necessary.

The sealant used between bolted panels may require replacement every seven to fifteen years. Consult the panel supplier or sealant manufacturer to determine this.

The bolted connections used with precast wall panels can corrode if they are exposed to the weather. However, they are typically positioned to be protected from exposure.

If the exterior surface is painted, it is normally repainted with a concrete paint approximately every seven years. The time to repainting is somewhat longer than with most wood surfaces because of the greater durability of paint on concrete.

Energy

Small buildings with insulated precast walls are believed to use 20 percent to 30 percent less energy compared with the same building constructed with standard wood frame walls. This will depend on the R-value of the panels used, however. Most systems offer the option of thicker foam to provide high R-values at little extra cost. The effectiveness of the insulation will also depend on the thermal breaks in it. Foam with an occasional composite plastic pin through it should provide nearly its full rate R-value. Many pieces of steel piercing the foam might reduce the R-value of the wall somewhat.

Air infiltration through the walls should be greatly reduced. There are few joints, and these are sealed with sealant or grout. The thermal mass of most precast panels is high.

Sustainability

Precast panels have the potential to have high-energy efficiency. The panels may enable a building to qualify for LEED points for energy efficiency.

They may also help to qualify for a credit for use of local materials. The panels usually come from within the 500-mile radius allowed by the LEED system.

The panels may or may not help a building qualify for a credit for recycled content. The concrete may contain some recycled material, such as fly ash. Rebar sometimes is made with recycled steel. It would be necessary to consult the manufacturers to determine either of these things.

Like many concrete products, precast walls are frequently cited as a method of reducing the use of trees for lumber.

Aesthetics

Because the panels have faces of exposed cast concrete that is created in a factory, precast walls have some unique options for their finishes.

Interior

Full sandwich walls and usually hollow core walls have a concrete surface inside. These can be plastered directly for a smooth finish. They may also be simply painted where uncovered concrete is an acceptable indoor finish.

Half sandwich panels show foam and studs toward the interior. In most cases this is covered with conventional gypsum wallboard.

Exterior

Precast walls panels can be finished with various products attached to the concrete on the site after they are erected. They may also have a variety of attractive, durable finishes cast in at the plant.

The simplest finish applied on the site is paint and nothing else. This leaves a flat concrete surface with any desired color. It is a highly economical finish.

Stucco may be troweled directly to the concrete. This is a more economical application than onto wood frame because it is unnecessary to apply a water resistant membrane or any form of lath or mesh for adhesion or strength. Stucco adheres well to plain concrete.

A thin brick veneer or other thin masonry veneer may be cast onto the face of the tilt panel at the plant. This can be very economical. It is also possible to add a conventional masonry veneer to create a double-wythe wall. However, this may be more expensive.

It is also possible to adhere various tile products to the exterior with special adhesives.

It is also possible to create a wide range of textures and fine finishes directly on the concrete of the exterior panel surface. Table 8-2 summarizes some of the more popular ones. These are virtually zero-maintenance because they

Table 8-2. Popular Finishes Cast into Concrete at the Plant.

Finish	Description
Colored concrete	Integral pigment in the concrete to provide almost any color permanently and without maintenance.
Exposed aggregate	Removal of the outer layer of cement paste to reveal the stones in the concrete, which may be selected for color and finish,
Form liners	Contoured and textured plastic surfaces set on the casting bed to give the concrete any of a wide range of finish surface shapes. May be painted later.
Thin brick	Half-inch thick clay brick embedded into the face of the concrete to create a finish of genuine brick.
Thin block	Two-inch thick concrete unit embedded into the face of the concrete to create a finish of genuine architectural masonry.

8-5 *Form liners in place to create a pattern on the wall surface.* Dukane Precast, Inc.

are usually created with permanent pigments and are formed of concrete. Finishes created on precast panels have the potential to be particularly high quality because of the control possible in the plant (Figure 8-5).

Key considerations

Careful advance planning is important to a successful precast building. With all features and dimensions planned out, panel installation and the rest of construction go smoothly. Any changes made after panels are constructed are more difficult and expensive. For some significant changes, panels have to be rebuilt at the factory and substituted for the original panels at the site.

The correct positioning of all panels is important. The manufacturer ordinarily provides a diagram showing where each panel is to be set. If there is an error in the field, by the end of installation, it will usually become clear that the remaining panels do not match the remaining spaces. It will be necessary to lift out some of the earlier panels and move them to their correct locations.

As with other concrete walls, in highly insulated precast buildings, HVAC contractors may tend to oversize the equipment. This can increase costs and the potential for equipment and humidity problems. The usual solutions apply: find contractors experienced with superinsulated or high-efficiency buildings; find contractors with sizing software, or get them to use the software available from the Portland Cement Association (www.concretehomes.com) or, in a larger project, rely on the mechanical engineer to calculate the HVAC needs accurately.

Precast walls are also relatively airtight. If the other parts of the building envelope are also tight, this may not produce sufficient air exchange. The HVAC contractor or engineer should be aware of this and provide for some mechanical ventilation to the outdoors.

Availability

Precast panels can be shipped from their manufacturing plants to almost anywhere. However, shipping long distances increases cost significantly. It is therefore important to locate precasting plants that provide the types of panels desired and are also reasonably close to the job site.

Lists of precasters are available from the web site of the Precast/Prestressed Concrete Institute (www.pci.org) and the National Precast Concrete Association (www.precast.org). Each one has a directory of member companies. Search these for makers of structural wall panels. Since many different types of panels are available, it may be necessary to inquire at each plant to determine which produces the types of panels that fill a particular need.

Crew availability

Most manufacturers selling panels for single-family homes also provide the crew to erect their panels. The general contractor arranges the timing and work of this crew with the manufacturer.

In larger buildings, the general contractor must usually arrange for the panel installation. In this case the panel crew may be separate from all the other trades. There are established precast installation subcontractors across North America. Many experienced commercial construction crews are also qualified and available to do the work. When a crew must be trained to do this work from scratch, it is usually sufficient to train one or two carpenters with good construction skills and have them supervise semi-skilled and unskilled workers. However, all workers must have safety training.

Support

Panel manufacturers maintain a staff to answer questions and visit job sites when needed. Some also have a network of local distributors and field representatives. These distributors and representatives are a useful source for technical support.

For certain broad questions it may be helpful to call the related trade associations, including the Precast/Prestressed Concrete Institute (www.pci.org) and the National Precast Concrete Association (www.precast.org) in the United States. The Canadian Precast/Prestressed Concrete Institute lists its members, products and systems, and offers technical literature on its web site (www.cpci.ca).

Current projects

It may be possible to locate projects to visit by contacting local precasters and asking what buildings they are currently supplying. Locate these by searching the on-line directories of the relevant trade associations.

9

Removable form walls

Removable form walls are cast in their final position in forms made of steel, aluminum, or wood. The forms are stripped to reveal a solid concrete wall. Although these walls are traditionally used mostly for basements and foundations, new systems have appeared that make them suitable for above-grade walls as well. These systems use forms that create all the common features and openings of above-grade walls. Some even include methods of casting a layer of foam insulation into the wall. The concrete, its steel reinforcing, and the foam may be varied to make these walls attractive for a wide range of above-grade building projects (Figure 9-1).

Unlike some other systems, removable forms concrete can also be practical for interior walls and components such as stairs. Some brands of formwork offer the needed parts to create these. With the formwork in position, the concrete can be cast to create everything at one time.

Removable forms have long provided economical, high-strength walls. Modern form equipment tailored to above-grade walls and attached insulation are recent developments. In the last five years the business has grown to several thousand small buildings constructed per year in North America. Growth is rapid.

Existing formwork crews can adapt to construction of these above-grade walls. This provides great potential labor availability. These crews also have most of the necessary forms. However, the formwork investment is large for a new crew. Constructing unusual wall features may require further investment in forms and accessories.

History

Walls cast in place with removable forms are one of the earliest types of structural concrete walls constructed. Concrete was cast into wooden forms to cre-

9-1 *Crew setting removable forms in position.*

ate walls as early as the 1850s, soon after the development of modern cement. In those days the forms were typically constructed of wooden posts and boards. The posts were staked into the ground around the building perimeter, along both the inside and outside lines of the walls. The boards were attached to the posts on each side to create two "walls" of boards that the concrete would be cast between. The boards ran horizontally from post to post. They had to be closely fitted to avoid gaps that the concrete might leak through. After the concrete was cast inside, the crew pulled the posts out of the ground and stripped the boards from the concrete.

Inventors made advances in the formwork over time. In one famous case, Thomas Edison created steel formwork for casting concrete houses. It was reusable, to avoid the need to "build two walls out of wood to create one concrete one." The forms even included details to create decorative features such as moldings. In 1904 his crews constructed twelve homes with his system in Union, New Jersey (Figure 9-2).

A pair of key developments became widespread after World War II. The first was to use wide panels stood vertically for the forms instead of posts and horizontal boards. Each panel covered a larger area. The vertical forms also eliminated the need to cut a large number of boards to length at the end of each wall. Another important development was the _snap tie_. Snap ties were steel rods that connected the inside and outside forms and held them a constant distance apart without the need for extensive outside bracing on either side. After the concrete cured, workers broke off the ends of the snap ties.

With steady growth, removable forms concrete became the most common method of creating basements. It had always been used in a few above-grade walls, but this was not common in small buildings.

9-2 _Crew casting concrete for one of Thomas Edison's concrete homes in Union, New Jersey._ Portland Cement Association

When above-grade concrete began to achieve new success in the 1990s, forms suppliers and inventors created new components to make traditional formwork suitable for above-grade work. Several innovative contractors began building the walls of houses with removable forms.

Rising interest led to the formation of the Concrete Homes Council (CHC). The CHC is a trade organization whose members are forms suppliers and other parties interested in providing information and promoting the use of above-grade removable forms walls for homes. Many of its members are the forms companies who have adapted their products for use above-grade. The Concrete Homes Council is a special division of the Concrete Foundations Association (CFA, www.cfawalls.org), a trade association representing the forms suppliers and others involved in the construction of removable form walls below-grade.

In 2004, an estimated 2,000 single-family homes in North America were constructed with above-grade walls cast with removable forms. The same forming systems were also being adapted to construct commercial buildings.

Market

Most of the single-family homes constructed with removable forms are currently built in Florida. Others are scattered throughout the Southeastern United States, and some other locations. The Florida homes are primarily the work of large builders. They are producing both custom homes and higher-end developments of dozens of houses. The houses in other parts of the country tend to be single-unit projects.

The same forming systems are also used for a variety of nonresidential buildings, as buyers and contractors experiment with them. However, most were inhabited buildings, or other types of buildings in which control of the interior environment is important. This includes multifamily buildings, offices, hotels and motels, retail stores, and refrigerated buildings. Some industrial buildings have also been constructed with removable form walls because of their durability.

Advantages to the owner

Removable form walls share all the major advantages of a reinforced concrete wall: strength; resistance to disasters, fire, mold, insects, and vermin; sound reduction; thermal mass and air tightness for energy efficiency; and durability.

They can also be outfitted with insulation to construct highly energy efficient buildings.

Formwork now exists to construct many interior features with concrete as well. This includes interior walls, staircases, and the like.

It may be important with removable forms to employ building designs that match the forms available with local crews. Acquiring the formwork to construct unusual wall features or odd building dimensions may take time and add to project cost.

Advantages for the contractor

Removable form walls can be a fast and efficient way for a contractor to begin constructing value-added concrete buildings. Most of the formwork and skills required are already in the possession of traditional basement crews. The construction method is also familiar to other building professionals. This can ease the task of finding designers and others to work on the projects. It can also make local acceptance easier.

For the concrete contractor, above-grade walls are a large opportunity to expand business. In many areas the volume of above-grade walls is several times the size of the local basement market.

If the formwork parts are not locally available for this new application, the investment required to get them could be sizable.

Components

Removable form walls consist of concrete and steel reinforcing bar.

Although not part of the final wall, the formwork is key to construction. Conventional forms today are panels that are two feet or three feet wide. They are available in different heights that match the height of the wall to be built. Most contractors have eight-foot and nine-foot forms. Some have ten-foot forms to build taller walls. These are typically made of wood or steel or aluminum. Each panel consists of a flat surface, backed by a frame to stiffen it.

For construction, pairs of panels are set face-to-face a constant distance apart. Successive pairs of panels are set side-by-side around the perimeter of the building in the position of the walls. Opposing panels are held a constant distance apart by snap ties. The ties extend through the forms at each joint between pairs of panels. They are held in position by locking pins or other small parts. The thickness of the form cavity may be adjusted by using ties of different lengths (Figure 9-3).

The new insulation systems for removable forms use sheets of plastic foam. Expanded polystyrene (EPS) and extruded polystyrene (XPS) are the types of foam most commonly used. It may be of different thicknesses, depending on the level of insulation desired. Typically the sheets are the same width and height as the form panels to fit between them neatly.

The other key components of the insulation systems are fittings that install the foam in position. These vary with the type of system: *face insulation*, or *center insulation*. In face insulation systems, the fittings are plastic parts. They clip over the edges of the foam sheet and fit into the forms to hold the foam flat against one of the two inside faces of the form panels.

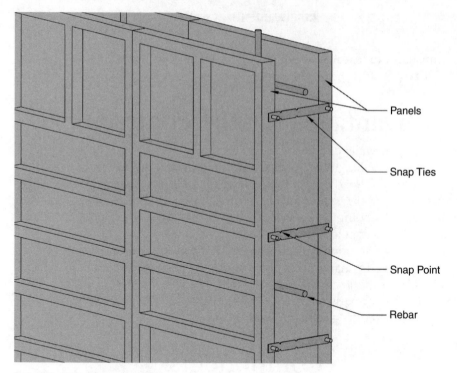

9-3 *Wall forms and rebar in place for producing an uninsulated wall (above) and with foam and fittings inserted between the form panels to produce a wall with face insulation (below).*

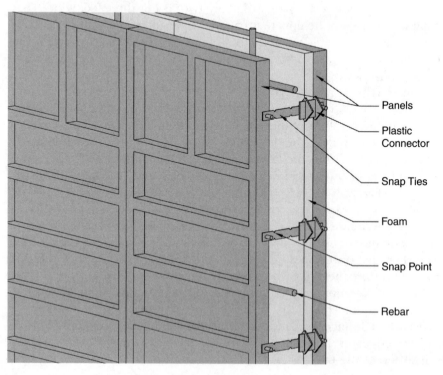

In center insulation systems, the fittings are composite pins that pierce the foam and extend out a few inches from either face. These hold the foam in the center of the cavity, away from either of the form panels.

A third option is *post-applied insulation* systems. With these, the insulation is installed after the concrete wall is complete and the forms are removed. They use concrete fasteners to attach either furring strips or a stud wall to the face of the bare concrete wall. A crew then installs fiberglass or other insulation between the strips.

Uninsulated wall assembly

If there is no insulation, the final wall is simply reinforced concrete. Typical thicknesses are 6 inches or 8 inches. However with the proper ties, wall thicknesses of 4 inches to 24 inches are possible.

There is typically one line of horizontal steel reinforcing bar around the wall perimeter about every four feet up the wall. There are usually vertical bars running up the wall as well. These are typically spaced every foot to every four feet along the wall, depending on the strength required. There is also one bar at every corner and bars along either each side of every door or window opening, plus horizontal bars above the openings.

There are also dowels extending out of the foundation below. These align with the vertical bars in the wall above and overlap them.

Insulated wall assembly

An insulated wall has a layer of foam. Different insulation systems handle the foam differently.

With a face insulation, the foam is positioned on one face of the wall. This is usually the inside, but it may be the outside instead if that is preferred. The foam is held securely to the concrete wall by plastic fittings. These have one end that extends into the concrete and is embedded there. The opposite end of the fittings extend along the edge of the foam sheet and have flanges that hold the foam against the concrete. The flanges will also accept a fastener such as a screw and hold it. They are useful locations for attaching finishes and fixtures to the wall after construction.

In center insulation systems, the wall consists of a layer of concrete on each face with a layer of foam in the middle, pierced by pins. The ends of these pins extend into the two layers of concrete and are embedded in them. Usually the inside layer of concrete is thicker, contains the rebar, and is designed to carry the structural loads of the building. The outside layer is thinner. It protects the foam and bears the exterior finish.

Post-applied insulation systems allow attachment of the foam after the concrete wall is complete. Vertical furring strips are attached to the face of the concrete with concrete nails or other mechanical fasteners. The strips may be

9-4 *Walls with post-applied insulation.*

spaced 16 inches apart to match the schedule of the studs of a frame wall. Between the strips is fiberglass or other insulation (Figure 9-4).

The window and door openings are lined with a buck, or formed with a blockout to leave plain concrete on all sides.

Electrical placement within walls

There are different possible locations for electrical boxes and cable. The cable may run through conduit cast into the concrete, with the boxes fitting into small blockouts in the concrete. This is common in walls with a concrete surface on the interior.

It is also possible to surface-mount metal pipe conduit and surface mount boxes onto the concrete wall, where this is acceptable.

On walls outfitted with foam insulation on the interior surface, the cable and boxes typically fit into chases and cutout in the foam. When the interior holds furring and insulation, the electrical typically runs in the bays between the furring, much as it does between conventional framing studs.

Cost

Removable form contractors typically charge about $6.00 to $7.50 per gross square foot for a completed insulated wall of average complexity.

The cost of the walls may be more than wood frame in a small building. However, the cost will be more nearly equal when high strengths are required. This is because frame walls typically require more expensive measures to raise their strength. This is often the situation in areas with high winds, where building codes require higher wall strengths. Higher strengths are also often required as the building gets larger, to sustain the greater loads of a larger building.

Table 9-1 includes a cost breakdown for typical removable forms walls with insulation.

The cost of removable forms walls increases with designs that have greater complexity or more unusual features. Large numbers of corners and openings add complexity and increase labor and cost. Features such as irregular corners, unusual elevations, and the like, are not all easily handled by standard formwork. Therefore workers may need to construct special forms to handle the job or pay to lease nonstandard forms.

Costs can decline significantly with repeated buildings or projects. Builders in Florida have constructed developments with a large number of similar homes. With such repetition, the contractor can gather the necessary formwork, then set it and reset it for successive buildings. The forms are used many times over, which spreads out their cost.

Table 9-1. Approximate Costs of Removable Forms Wall Construction.

Item		Cost per Square Foot*
Labor		$2.00
Materials		$1.95
Concrete	$1.25	
Rebar	$0.35	
Miscellaneous	$0.35	
Pump rental		$0.35
Foam insulation		$2.00
Total		$6.30

*All costs in U.S. dollars per gross square foot wall area.

Code and regulatory

Removable form walls are covered under the general concrete provisions of the International Building Code (IBC). These are based on the recommended engineering of the American Concrete Institute (ACI). That is contained in ACI's "Building Code Requirements for Reinforced Concrete", a document also referred to as ACI 318. On multifamily and nonresidential buildings, custom engineering by the designers is standard, anyway. This design must be in accordance with the IBC concrete provisions.

Removable form walls for basements and foundations are also covered by the International Residential Code (IRC) in Section R404. Their use in above-grade walls for houses is covered by Section 611. Their use in above-grade walls is less familiar to many building departments, however. Some departments may request more information or even analysis by a structural engineer.

The foam used as insulation must meet the general fire resistance requirements of foam used in construction. The foam suppliers are responsible for meeting these requirements and having documentation to verify this.

Installation

Different construction methods have evolved for basement and above-grade wall construction. Including insulation also changes the installation.

The typical crew for a single-family home is one experienced supervisor and one or two laborers. On a larger building, there may be multiple teams of three or four workers.

Uninsulated foundation walls

Foundations are typically set on a concrete footing. This footing is normally constructed with a series of dowels extending out of it. These are positioned to stand approximately along the centerline of the walls to be constructed above. The crew lays out the wall design with lines on the footing. These mark the inside and outside edges of the wall to be built and the locations of openings.

Bucks or frames for windows and doors are usually all assembled before setting forms. Many manufacturers offer windows or frames for basements that are already constructed and ready to be set in position.

Next the crew sets the forms along the wall lines. Usually the forms for the corners go in first. These may be braced to prevent them from leaning out. Then the crew sets the forms next to the corners and works toward the center of the walls. If there are multiple teams on the crew, they can set forms along different parts of the perimeter at the same time.

Setting the forms involves standing one panel along the outside wall line, setting a matching panel along the inside line, and connecting them with ties along their edges.

The crew may install the rebar within the wall as they go. Typically one vertical bar is wired to each dowel The horizontal bars are set at required height and may be wired to the vertical bars or ties. The vertical bars are also wired into position at the top of the wall to hold them near the center of the form cavity (Figure 9-5).

9-5 *Forms with rebar ready for concrete.*

At opening locations, the crew sets the correct door or window frame be-tween the form panels. It is held in its correct location with nails or special fittings. The depth of the frame matches the thickness of the form cavity so that it fits snugly against the inside faces of the form panels.

Braces are installed to hold the forms plumb. Usually these go on the in-side, extending from a high point on the walls diagonally down to the inte-rior ground or slab.

When electrical is to be installed inside the concrete walls, the crew also puts conduit for the cable and blockouts for the boxes inside the forms.

Uninsulated above-grade walls

Above-grade walls could be constructed with the same methods as foundation walls. However, contractors have developed some alternative methods.

The biggest difference is that they often choose to set the rebar before the forms. They construct a rebar *cage* that stands above the foundation. The bars are all in their intended position and attached to one another, in something resembling a wire framework. The crew then sets the forms around the rebar. They wire the bars to the form ties as needed (Figure 9-6).

9-6 *Rebar cage ready for forms to be set around it.*

The openings are also handled differently. There are a wide variety of doors and windows available, and most are not designed to be inserted in removable form walls. Typical residential windows and some doors are designed to fit into an opening after the wall is complete. They are installed by running fasteners through the window frame into the walls immediately around the opening. Therefore the contractor installs a buck in the position of the opening, to hold the concrete out of the space and to receive the fasteners when the window or door is installed.

The width of the lumber for the buck is selected to match the width of the form cavity. At opening locations, the corresponding bucks are set in position. The buck fits snugly between the form panels. It may be fixed in position with nails or fittings embedded in the concrete that keep it from shifting.

Especially in commercial buildings, openings may be designed to have no permanent buck. Instead they will simply have concrete sides. In these cases, the "buck" is really a blockout that will be removed later. However, it is installed in the same fashion.

Insulated walls

With face or center insulation systems, the crew installs the insulation as it sets the forms. One sheet of foam goes between each pair of panels. It has the same height and length as the panels. Face insulation is set flat against either the inside or outside panel. The system's plastic fittings clip over the edge of the foam and onto the forms to hold the foam in position.

With center insulation systems, workers slide the foam between the panels. The pins through the foam protrude in both directions to keep the foam spaced away from each form panel.

If the insulation is instead to be post applied, no foam or fittings are installed during construction of the concrete wall. Later, after the forms are stripped, workers attach furring strips to the inside wall surface, spaced at regular intervals. Between the strips they hang fiberglass or some other insulation.

When the walls are outfitted with insulation on the interior surface, typically no conduit or electrical box blockouts need to go inside the form cavity. The electrical will run through the insulation layer.

Interior walls

If the design calls for interior walls constructed of concrete, then the forms are set at the same time as the exterior wall forms. Interior concrete walls are most common over a slab-on-grade foundation. This is better able to support the weight than most floor decks.

The formwork for the interior walls ties into the exterior formwork. It often produces a narrower wall because less strength is required. Interior walls are rarely insulated.

Placing concrete

Before placing concrete, the crew may go through a checklist to make sure all of the formwork is in order. Concrete delivery is typically scheduled for the next morning, so it can be called off if they discover anything wrong.

Typically the crew sets scaffolding along the wall all around the perimeter. This gives good access to all points while placing the concrete.

To place concrete into the forms, some type of equipment is needed to move the concrete. For a foundation with good access all around the perimeter, it may be possible to use the chute from the ready-mix truck delivering the concrete. However, for many foundations and all above-grade walls, some form of pump or other concrete lifting equipment is needed. There will be rental charges, but special equipment can be fast, precise, and does not require close access to all points of the wall.

The placement may proceed in a series of *lifts*. These are partial fills of the wall, usually about four feet high. However, depending on the skill of the crew and the strength of the formwork they may fill it up completely on the first pass.

Placement begins at some convenient point along the form wall. The crew moves around the perimeter. Trailing the person guiding the concrete into the wall is a second worker with a vibrator. This worker drops the head into the concrete about every foot along the wall. This proceeds until the crew goes entirely around. If the placement goes in lifts, they repeat the procedure until the wall is filled. If no further stories of concrete walls will go on top of the current one, workers trowel the top of the wall smooth.

During the placement, workers check the plumb of the wall and adjust the bracing as needed.

Form stripping

The forms are left in position while the concrete cures sufficiently. After this time, the crew returns to remove the forms. They break off the ends of the snap ties. This releases the form panels, allowing the crew to pull them off.

The openings contain a buck or blockout. When the openings are supposed to be solid concrete without a buck, the crew removes the blockout as well.

Upper stories and tall walls

In a multi-story building, the floor deck may be installed before the next story up is constructed. The crew then works off the floor. If the floor is not installed until later, the crew works from scaffolding. They set the forms on top of the walls below and proceed as they did in constructing the first story.

In a multi-story building the vertical rebar of the lower stories should be sized so that it extends past the top of the story a certain distance, usually about two feet. The vertical rebar in the story above are then installed at the same locations and overlap the bars from below. This ties the two stories together.

Connections

Methods of connecting concrete floors and roofs to removable form walls are covered in the chapters on those floors and roofs. Here are the methods of connecting frame and heavy steel floors and roofs.

Floors

Wood and light-gauge steel floors are typically attached by means of a ledger. The ledger may be connected to the wall with anchor bolts adhered into holes in the concrete after the forms are stripped. This avoids the necessity of putting holes through the forms to hold anchor bolts in position to cast them into the concrete.

For heavy steel floors, the crew installs weld plates inside the formwork to embed in the concrete. Later workers may weld the steel joists or trusses to the plates. As an alternative, they may also weld angle iron to the plates to create a shelf on which to set the joists or trusses.

Roof

One of the most common methods of attaching a frame roof is by means of a top plate. It is also possible to attach each roof truss or rafter to the wall directly by means of steel "hurricane straps." Heavier steel roofing trusses are typically welded to plates in the wall. The weld plates are cast into the concrete walls at precisely measured locations.

Other connections

Interior walls may be attached to the precast walls in various ways. If the exterior concrete walls are not insulated on the inside, interior walls may be attached with common concrete connectors, such as concrete nails or concrete screws. If the concrete walls are insulated on the inside, interior walls can be nailed or screwed directly to plastic fittings over the insulation (in the case of face insulation) or the furring strips (in the case of post-applied insulation).

Lightweight fixtures can be attached in the same ways as interior walls. Heavy fixtures may require heavier concrete connectors, such as adhered anchor bolts.

Architects

Architects normally adapt readily to designing with removable forms concrete. It follows rules similar to those of other methods of construction they are likely to be familiar with. The system is also well documented in the general construction literature.

It is useful for the designer to learn what types of forms are available from the crews that will be performing the construction. If the design includes wall dimensions or features for which the forms are not readily available, it may be necessary to ship in and rent special forms at additional time and expense.

Engineers

Nearly all structural engineers have extensive training in removable forms concrete. The design rules for it are extremely well documented. To find these the engineer can refer to the concrete section of the IBC or to ACI 318.

Many engineers have not used this form of construction for the above-grade walls of small buildings. In this case they may be inclined to specify greater wall thicknesses and the use of more rebar than is strictly necessary. This is conservative. It adds greater safety to their design in a situation where the engineer is uncertain. However, it may add unnecessary cost and complexity to construction. To avoid this, it may be wise to search for an engineer experienced with small buildings or one who is flexible and willing to spend some extra time coming up to speed on the system. It may be necessary to pay something extra for the engineering of the first few buildings, until the engineer becomes skilled at designing this type of building.

Training

There are already thousands of crews across North America installing concrete walls with removable forms. It is easiest to employ such a crew, or at least get your supervisors from among the experienced workers of one of these crews.

There are some new steps to learn for forming and insulating above-grade walls of small buildings. Training materials are available from the manufacturers of the forms and insulation systems designed for this form of construction. These materials are mostly manuals. Site visits may also be available from the manufacturers.

Maintenance and repair

Removable forms walls are highly durable. The exterior surface is usually concrete. The interior surface may be concrete or insulation and studs covered with gypsum wallboard. There is generally no routine maintenance, and repairs are rarely necessary.

The concrete is impervious to rot, rust, and mold. It is also highly resistant to fire. However, this does not mean that the occupants can relax their fire precautions. Most fire injuries occur as a result of smoke inhalation. The smoke comes from the burning of the contents of the building, which arises regardless of what the walls are built with.

Depending on what exterior finish is chosen for the walls, it may require some periodic maintenance.

Energy

Based on experience with other concrete walls, small buildings with well-insulated removable form walls should use 20 percent to 30 percent less energy compared with the same building constructed with standard wood frame walls. Many insulation systems allow the user to choose the R-value by varying the type and thickness of the foam used.

Air infiltration through the walls should be among the lowest of any wall system. There are typically no joints. The thermal mass is also high. The thermal mass evens out extreme temperatures, lowering heating and cooling requirements.

One advantage of removable forms walls for energy purposes is the ability to place the insulation on the outside. Where to locate the insulation is usually determined by such factors as the convenience of attaching finishes and installing electrical wiring. However, according to engineering simulations, insulation located to the outside results in somewhat greater energy savings from the thermal mass effect. In addition, the mass of the walls can store substantial amounts of heat inside. This is valuable for passive solar heating schemes.

Sustainability

The high-energy efficiency of insulated removable form walls is considered a significant environmental advantage. It may also help to qualify the building for LEED points for energy efficiency.

The walls may also help qualify for a credit for use of local materials. The concrete almost always comes from within the allowed 500-mile radius. Often the insulation does as well.

The concrete used may or may not help a building qualify for a credit for recycled content. It may contain some recycled material, such as fly ash. Rebar sometimes is made with recycled steel. It would be necessary to consult the manufacturers to determine either of these things.

Like many concrete products, removable form walls are frequently cited as a method of reducing the use of trees for lumber.

Interior aesthetics

On a wall fitted with insulation on the interior face, the finish inside is almost always gypsum wallboard. This is screwed to the insulation system's plastic fittings or furring strips.

Other walls have a concrete interior surface. These include uninsulated walls, center-insulated walls, and walls with the insulation installed on the outside. The interior surface can be plastered directly for a smooth finish. It may also be simply painted where uncovered concrete is an acceptable indoor finish.

Exterior aesthetics

Removable form wall panels have a wide variety of exterior finish options. This is partly because the exterior surface may be exposed concrete. This makes it possible to use novel concrete finish methods, in addition to traditional finishes.

Stucco may be troweled directly to the concrete. This is a more economical application than onto wood frame because it is unnecessary to apply a weather-resistant membrane or any form of lath or mesh for adhesion or strength. Stucco adheres well to plain concrete. Stucco over a concrete surface is also strong and impact resistant. Workers can give the stucco a wide range of surface textures. It is also possible to adhere trim pieces made of foam or other light material to the concrete, then stucco over these. This produces various types of surface relief that may be attractive. The stucco may be integrally colored, or may be painted after it is applied.

A brick veneer or other masonry veneer may be added in the conventional way. Brick ties are either cast into the concrete wall, or attached with concrete fasteners after it is up. The brick is stacked in front, with the ties embedded into the mortar. This type of finish produces a double-wythe cavity wall.

The panels may be cast with a wide range of surface textures and shapes, This can be done with special textured forms or by use of *form liners*. Form liners are flexible mats affixed to the inside of the form panels. The concrete cast into the wall is molded with an exterior surface that matches the surface of the special forms or form liner. Forms and liners are available to create a wide range of desirable and attractive finishes on concrete. Some resemble traditional finishes. Form liners can give concrete a surface that is shaped and textured like the wall surface of brick or wood siding or stone. They may also produce the shapes of moldings or other wall features. Yet the entire surface is durable concrete. These features may be painted as desired after the wall is up. Most of these methods have slight seams at the edges of the panels, which might be figured into the design or touched up after stripping forms.

If insulation is placed on the outside of the wall, virtually every finish used over frame construction can be attached in nearly the same way. This includes sidings of vinyl, fiber-cement, wood, and hardboard, in addition to stucco and brick. The fasteners connect to the plastic fittings of the insulation instead of to the studs or sheathing.

The finishes normally nailed or screwed to frame walls can also be attached to a concrete wall that is not outfitted with insulation on the outside. First workers attach furring strips to the surface with concrete fasteners. Usually they are installed in the same pattern as the studs of a frame wall. The finishes then connect to the furring. This will add the cost of the furring to the project, making the finish work somewhat more expensive than it would be over a frame wall. However, if properly constructed it also creates a drainage plain in front of the wall.

Key considerations

Significant construction problems with experienced crews are rare. The most common one is voids in the concrete. These are easily prevented with proper consolidation. Most voids are also visible after the forms are stripped, and may be repaired by filling them with a patching compound. Normally the forms contractor is responsible for making this repair.

Especially with older forming systems, corners may sometimes not be precise at the edge. This is not a concern below grade, but above it may be necessary to grind down any imperfections or cover with the finish material to get an attractive appearance.

As with some other systems, in highly insulated removable form buildings, HVAC contractors frequently install equipment that is larger than required. The equipment costs more than necessary, robbing the owner of possible savings. In addition, oversized equipment may not run efficiently. As discussed in Chapter 5, one method to prevent the selection of oversized equipment is to find an HVAC contractor who uses modern software or who has experience with superinsulated buildings. These contractors are more likely to do more accurate calculations than those who use rules of thumb based on building size. This is less of a problem in larger projects. In these projects, normally, a professional HVAC engineer designs the system.

The use of removable form walls may also lead to a very tight building with low infiltration. If the other parts of the building envelope are also tight, this may not produce sufficient air exchange with the outdoors. The HVAC contractor or engineer should be aware of this and provide for some mechanical ventilation to the outdoors.

Equipment and crew availability

Although crews may be imported, it is usually most economical to use local ones. There are thousands of removable form crews across North America. However, not all of these have experience with above-grade walls for small buildings, or have all the appropriate forms.

One method of locating a suitable crew is to contact the suppliers of the forms. Most of these are members of the Concrete Homes Council of the Concrete Foundations Association (www.cfawalls.org).

Support

The forms manufacturers maintain a staff to answer questions and visit job sites when needed. Some also have a network of local distributors and field representatives.

For certain broad questions it may be helpful to contact the trade association serving this construction method, the Concrete Foundations Association (www.cfawalls.org).

Current projects

It may be possible to locate projects to visit by contacting local contractors and form company representatives. You can find them with the methods listed in the preceding sections.

10

Tilt-up walls

In tilt-up construction, concrete walls are cast on the ground in the horizontal position. They are then lifted into place with a crane and structurally connected to the roof and possibly to the foundation to create the building shell (Figure 10-1).

Tilt-up's strong suit has long been the rapid and economical creation of large, durable buildings with a simple rectangular floor plan. Tilt-up has taken much of the North American market for "big box" buildings. They include warehouses, distribution centers, large factories, and warehouse stores.

But rapid innovation over the last ten years has made tilt-up competitive in a wide range of small and premium buildings as well. They are available in almost any size, with premium finishes and low or high levels of insulation. They can be made with many traditional shapes or with distinctive arrangements of panels that are difficult to replicate with any other construction method.

Tilt-up produces nearly seamless concrete walls with only a few joints that are easily controlled. It is relatively fast and requires less labor and fewer skilled workers than many other wall construction methods. However, when constructing small buildings, a high degree of precise planning is critical to project success.

The Tilt-up Concrete Association estimates that in 2004 nearly 300 million square feet of tilt-up walls were produced in North America. They enclosed about 600 million square feet of building space. These figures have more than doubled from their levels of ten years ago. According to reasonable estimates, small tilt-up buildings account for somewhere in the tens of millions of square feet per year. The small building segment appears to be growing even more rapidly than the tilt-up industry as a whole.

History

Historians credit the Amish with the practice of building their walls on the ground and hoisting them into position. They lifted with a series of ropes and pulleys. The method allowed them to do most of the assembly work on the ground, which made it easier and more efficient.

10-1 *Tilt-up wall panel being lifted into position.* Tilt-Up Concrete
Association and Seretta Construction

As early as 1905, an officer in the Army Corps of Engineers named Robert
Aiken conceived of a plan to cast concrete walls horizontally at the job site. The
walls were cast next to their final intended positions and tilted into place with a
derrick. Over the next two decades, Aiken supervised the construction of military
buildings with the method, including target abutments, barracks, and gun houses.

Tilt-up construction got a big boost with the development of mobile cranes
during World War II. These provided the means to lift the heavy panels efficiently and economically. After the War, commercial contractors in southern
California adopted the method of casting concrete walls horizontally on the job

site and then lifting them. As they refined it, tilt-up became efficient for the construction of buildings with a large, rectangular footprint. It grew in California and then spread across North America and to other countries.

In 1980 the American Concrete Institute (ACI) formed ACI Committee 551 to study tilt-up construction and make recommendations for its practice. In 1986, industry contractors, manufacturers, and professionals formed the Tilt-up Concrete Association. This is the trade association representing tilt-up. It promotes the acceptance and quality of tilt-up construction.

The industry's move to smaller and more architectural buildings has been rapid. A few specific advances have been especially important to this. Inventive contractors have developed methods of producing panels for a small building, and doing it economically. For one, innovations in forming have made it practical to cast all the wall panels on-site even though the space available on the floor slab is limited. Vendors have created a wide range of appealing exterior finishes to suit every application. Forming techniques have advanced to make unusual and interesting shapes feasible. Finally, inventors have created quality systems for insulation.

Market

Tilt-up construction is used widely throughout the United States and parts of Canada for large commercial buildings. Historically it is most used in the southern half of the United States. California has been the state with the highest volume of tilt-up construction ever since the 1940s. Today Florida and Texas are the next biggest tilt-up producers. However, many central states now have large tilt-up construction rates as well, and it has become popular in such northern areas as the Pacific Northwest, British Columbia, the lower Midwest, Pennsylvania, New Jersey, and Atlantic Canada. This shift north has occurred as contractors have developed ways to perform tilt-up work in cold weather.

Tilt-up may now be the most widely used method of wall construction for big box buildings. Among smaller and more architectural buildings, it has grown rapidly for use in schools, retail buildings, community and civic buildings, and religious buildings. The advances in construction techniques for small buildings, in premium finishes, and in insulation systems have made it possible for tilt-up to appeal to these markets. It appears that rapid growth will continue.

Advantages to the owner

Tilt-up construction offers the usual advantages of concrete walls. These include strength, durability, sound reduction, and resistance to disasters, water, and insects. They also include air tightness and thermal mass.

Traditional tilt-up also offers a short construction time at a reasonable cost. This is where it has excelled in the past. This benefit also has a strong appeal in certain types of smaller buildings. Opening schools by September is often critical. Opening a retail store a month early may bring in hundreds of thousands of dollars more in revenue.

Tilt-up offers some flexibility. Since it consists of large, reinforced panels, openings can be located almost anywhere in the field of the wall.

More recently, tilt-up has improved on other dimensions as well. For one, there is more design flexibility. Many buildings are constructed with finishes that match the look of traditional finishes like stucco, brick, stone, and architectural block. Traditionally always flat, panels built with new methods are sometimes curved or formed with depth relief. Low to high levels of insulation is an affordable option as well.

Advantages for the contractor

For the contractor, tilt-up can be highly efficient. Wall construction is fast. Once cured, the panels can generally be lifted in one day. This lowers the risk of delay from things like the weather or labor availability. It also allows the other phases of work to move in quickly. Most tilt-up work is done on the ground, where it is easy to inspect and manage. The usual work sequence is to create the floor slab first. This provides a stable work platform, which also makes work efficient and less sensitive to weather. In addition, only a few highly skilled supervisors are necessary on most projects. This provides flexibility to bring in extra workers and get them to be productive on short notice if necessary.

To achieve these efficiency levels, planning and thoughtful supervision are necessary. Errors that require extensive rework or require the lifting crane to come out more than once can quickly increase expenses. This is especially true for small buildings. The methods developed for large buildings can be adapted for smaller ones and still be efficient. However, mistakes in sequencing or layout can have an even bigger effect on the smaller budget.

Components

Tilt-up construction uses mostly commodity materials. The concrete is conventional. The exact mix used may be formulated to produce high strength or to gain strength quickly. This allows for safely lifting the panels soon after the concrete is cast.

The steel reinforcement is conventional. It is set on chairs. Most often used for the forms is ordinary 2× lumber. However, systems of metal or plastic forms that are reusable are also available. They are held in position on the slab with small diagonal braces sometimes called *angle supports* (Figure 10-2).

Also installed in the panels are special steel hooks or brackets called *lift inserts*. These stand in the casting bed and are wired into position before concrete is cast. After casting, they are exposed above the concrete so that the crane can hook onto them for lifting the panel. The suppliers of the lift inserts design, test, and provide engineering for them.

10-2 *Formwork for a wall panel.*

Weld plates and other connection hardware are sometimes installed as well. These are referred to as *embeds*. They are set and wired into position much like lift inserts.

Most insulation systems for tilt-up consist of foam sheets with composite or plastic connectors that extend through the foam. The connectors have deformed ends to embed securely in the concrete on either side. These are also designed and engineered by their suppliers. The foam is usually extruded polystyrene (XPS). Most commonly used is two-inch thick insulation foam. This adds an R-value of about ten to the wall. However, material as thick as 6 inches is available, which would add about thirty to the wall R-value.

Some items commonly used for decorative purposes are strips of wood or plastic called *rustication* or *reveal strips.* These are fastened to the floor slab where the panels are cast. They leave indented lines on the panels that accent it or divide it into different areas that can be painted different colors or treated with different finishes. They are frequently ordinary construction lumber or wood molding that is cut to the desired profile. There are also plastic extrusions designed to do the job. They come in a wide variety of profiles.

When the panels are lifted, it is necessary to hold them in position until the rest of the structure is connected to them. This is done with diagonal *braces* made of tubular steel. They have fittings on each end to connect to the back of the panel and to the floor slab. These are provided by the same companies

10-3 *Braced wall panels.* Tilt-Up Concrete Association and CON\STEEL Tilt-Up Systems

that supply the lift inserts. As with the inserts, the supplier has designed and tested the braces, conducts the engineering analysis for the bracing requirements of each project, and provides the needed braces (Figure 10-3).

Uninsulated panel assembly

The standard tilt-up wall panel is usually about 7 to 12 inches thick. The thickness is determined by the strength required.

The amount, size, and location of the rebar are also determined by the engineering requirements. Normally there are vertical and horizontal lines of bar.

The verticals may be as close as every foot on center across the wall panel. If openings are positioned so that the panel has very narrow sections, vertical bars may be positioned even closer to reinforce these weak points. Horizontal rows of bars are typically located about every 16 inches up the wall. However, if the openings create narrow horizontal sections, there may be more bars in these.

There are usually also vertical bars along either side of an opening, and extra horizontal bars over the top. The concrete above a wider opening will have to span a greater distance, so it requires more reinforcement.

When the panels are in place in their vertical position, there is a space of one-half to one inch between adjacent panels. These joints are filled with sealant along the outside face of the wall.

Insulated panel assembly

The typical insulated panel has a layer of foam sandwiched between two layers of concrete. The outer concrete layer is two to three inches thick. It is typically reinforced with a layer of welded wire mesh. The inner layer of concrete is the structural layer. It is similar in thickness and in reinforcement to the typical uninsulated panel.

Extending through the insulation and into each concrete layer are the connectors of the insulation system. Their ends are embedded in the concrete, connecting the two layers. Standard spacing for the connectors is one every 16 inches vertically and horizontally. However, they may be spaced at different distances depending on the strength requirements of the particular situation and the system used.

Cost

Tilt-up subcontractors currently charge about $6 to $11 per square foot for uninsulated, unfinished tilt-up walls. This includes erection of the panels, sealing the joints, and correcting any surface imperfections on the panels. Adding insulation to the wall panels typically adds $3 to $4 per square foot to the cost.

The costs of construction for a typical medium-sized project in 2004 were approximately as in Table 10-1.

Table 10-1. Approximate Costs of Tilt-Up Wall Construction.

Item		Cost per Square Foot*
Labor		$2.45
Materials		$3.05
Forms	$0.25	
Concrete	$1.45	
Rebar	$0.50	
Cleaning and patching products	$0.25	
Inserts and embeds	$0.35	
Miscellaneous	$0.25	
Equipment		$1.70
Crane	$1.10	
Concrete pump, lifts, misc.	$0.60	
Total		$7.20

*All costs in U.S. dollars per gross square foot wall area.
Source: Tilt-up Concrete Association

The cost of tilt-up construction is sensitive to the size of the panels and the project. The cost per square foot goes down as the panels get taller, to about fifty feet in height. This is because it is not much more time consuming to create a large panel on the ground and lift it. Tilt-up walls are also usually less per square foot for buildings with large footprints. Economies of scale set in as the contractor can organize the production of a large number of panels into something resembling a production line.

More recently the cost of producing small buildings has become competitive in some situations. Projects as small as 5,000 square feet have won competitive bids in some cases.

As with most construction methods, the complexity of the floor plan and the wall shapes increases construction costs.

On a small building, the degree of planning can also have a large influence on costs. The construction of small tilt-up buildings involves more skilled labor. If the workers are not kept busy and used effectively, the panel forming costs increase. Similarly, crane rental is a large expense on a small building. Operations should be scheduled to lift all panels in quick succession. If instead the crane is held idle on the job site for extended periods, the rental cost increases sharply.

Code and regulatory

Tilt-up construction is covered by the reinforced concrete section of the International Building Code. It is not covered in the International Residential code.

Almost all tilt-up buildings are designed structurally by licensed engineers who submit their calculations and stamp the plans. Presented with this documentation, building officials readily accept plans for using tilt-up. In areas where it is common there will rarely be any questions unless something quite unusual is to be attempted. In areas where tilt-up is new or rare, the building department may ask some additional questions.

Installation

Small tilt-up buildings have become economical with the development of new methods that are different from traditional tilt-up construction. However, some aspects of traditional methods are still used.

Traditional panel forming

Before beginning work for the panels, it is necessary to prepare the foundation and floor slab.

The foundation of a tilt-up building is usually a continuous footing. The perimeter is excavated to cast the footing below grade. The floor slab is cast next, generally as level as possible. The slab will be used for casting the panels and possibly supporting the crane as well. Therefore the slab specifications may be greater than with other building methods.

In most cases the walls are all formed on the slab. If there is not enough room, it may be necessary to create an additional, temporary slab nearby to build some of the panels. This is important for having all the panels finished and ready to be lifted at one time. Creating some of the panels, lifting them, and then creating and lifting the rest is expensive because it requires leaving the crane idle for a time or bringing it out to the job site twice.

Plans generally include a layout for the panel positions on the slab. This locates the panels in a way that makes erection efficient. Typically each panel is built near its final position and panels that will be adjacent in the final building are formed adjacent to one another.

The crew snaps chalk lines on the slab for the outer dimensions of all panels. They select forms with a width equal to the design thickness of the panels. The forms go along the chalk lines, on edge. They are held in position with angle supports.

The forms for openings are installed in much the same way. Any rustication strips are also fixed to the slab at this point.

With the forms in place the crew sprays a coating of _bond breaker_ onto the slab. This material prevents the cast concrete from sticking to the floor slab.

Next, reinforcement, inserts, and embeds are installed. Their positions are given on plans. The rebar rests on chairs set at intervals of a foot or two along the reinforcement lines. The bars may be wired to the chairs. They are wired to one another where they cross. Most lift inserts are designed to rest on their own built-in legs that hold them at the correct elevation. Weld plates may be wired to the rebar, adhered to the slab, or attached to the side forms, depending on where they are located.

Concrete is cast into all forms in predetermined sequence. It may be placed directly from the chute of the concrete truck. However, some contractors prefer the greater range and control they get with a concrete pump.

As the concrete enters the forms for a panel, the crew rakes or otherwise moves it into position. One worker pulls the head of a vibrator through the wet concrete to consolidate it. Others trowel it smooth.

The time needed to cure the panels so that they will be strong enough to lift is worked out in advance. The sequence of the lift is also worked out in advance for maximum safety and efficiency.

First the crew locates and cleans off all lift inserts. They attach one or more braces to the back of the panels so they will be ready to brace to the slab as soon as they are in place.

Moving the crane takes time, so plans call for it to be located in the smallest number of locations for raising all the panels. The crane is outfitted with rigging that includes one cable in the correct location for each lift insert. Each cable has a hook-like mechanism that matches the lift inserts used. The crane operator drops the rigging over a panel. The crew hooks the lines to the inserts and moves clear of the lift. The crane lifts the panel into position.

The crew uses pry bars to shift the panel precisely to align with predrawn chalk lines on the footing. The panel may also have embeds that match up with dowels or other inserts in the foundation. These may be connected by various means to prevent the walls from sliding on the footing. The panel typically rests on shims that are adjusted to level the panel precisely. They also secure the free ends of the braces to the slab and adjust to bring the panel to precise plumb.

This lifting procedure is repeated for the rest of the panels.

The gap beneath the panels is filled with grout as soon as possible to provide a continuous bearing surface. The braces stay in position until the roof, deck, and other permanent structural members are attached to the walls. The braces cannot be removed until the engineer approves their removal.

Other workers elevated by lifts go over the panels to fill in or grind down any unacceptable surface imperfections. They fill the vertical joints between panels with sealant.

Forming panels for small buildings

With a small building, the wall panels typically will not all fit on the floor slab. In fact, the area of the walls may be several times that of the floor. To avoid the expense of extensive temporary casting slabs, the contractor uses a procedure known as *stack casting*. In stack casting, panels are constructed one on top of the other. There may be stacks that are four panels high. However, more than that is not usually recommended (Figure 10-4).

Deciding the position of all panels is where much of the planning of the construction process enters. In some regards it is like a puzzle. It is critical that the casting position of each panel be determined before the first one is formed. It is ideal to form each panel on top of another panel with exactly the same length and width. Then the forms of the upper panel can rest directly on the forms of the panel below. The upper forms will be braced with kickers to the slab. Where stacking panels of identical dimensions is not possible, smaller panels should rest on larger ones. Panels with openings should go on top of a stack. When it is necessary for panels with openings to be below the top, they will need to be covered with a layer of plywood or similar material before forming the next panel above.

It is also desirable to arrange the panels to help make the lift as efficient as possible. Ideally, the stacks will be arranged so the crane can easily pick up the panels in the order in which they are to be arranged around the building, and without the necessity of moving the crane often.

The bottom panels are formed up and built together according to the same methods used in traditional tilt-up construction. After they cure, the crew forms the second layer of panels on top of the first, with a coating of bond breaker sprayed on the backs of the first layer of panels. The second layer is then cast, and the process repeats until the top panels are done.

10-4 *Stack cast panels ready for lifting.* Tilt-Up Concrete Association

Lifting and securing the wall panels is done according to the conventional methods. For the early panels, the slab still may restrict crane and crew movements.

Insulated panels

Where panel insulation is used, it is measured, cut, and checked before any concrete is placed. The foam sheets and connectors are onhand and ready for each panel.

The outer layer of concrete is cast first. It is typically two or three inches thick. Welded wire mesh may be installed in it to add strength and control cracking. While this concrete is still pliable, the crew lays the foam on top and inserts the connectors through. They walk with small steps on the foam alongside the connectors or agitate the connectors to insure good concrete consolidation around them.

After the bottom layer of concrete is stiff the crew installs rebar, inserts, and embeds for the structural layer. The concrete is placed on top by the conventional methods.

Panel lifting and installation proceed as they do for conventional panels. When the panels are in position, a line of expanding foam adhesive may go in each joint before the sealant. This adhesive bridges the gap between the foam along the edge of one panel and the foam along the edge of its neighbor. It creates a continuous insulated layer around the walls of the building.

Finishes

Some popular tilt-up finishes are created with special steps performed during panel construction.

Colored concrete is sometimes used to create a novel, durable colored finish. Conventional concrete pigments are mixed into the concrete cast into the panels. To reduce costs, a layer of colored concrete only an inch or two thick may be cast first. Uncolored concrete is then cast on back of this to create the rest of the panel.

Exposed aggregate finishes are generally created with a chemical called a *retarder*. A retarder slows the curing of concrete. The crew sprays a retarder onto the casting slab instead of a bond breaker. The concrete on the surface of the panel remains pliable as a result. After the panel is in position, workers treat the surface with a *power wash*. This is water shot out of a nozzle under high pressure. It forces the soft cement off the surface, exposing the stones in the concrete. An alternative method of exposing aggregate is by sandblasting the panel surface lightly after it is lifted.

Form liners are thin sheets or mats of plastic or rubber-like material with desirable surface texture or details. Liners are installed over the casting bed before rebar or inserts. The concrete cast onto the liners takes their surface shape to create the desired appearance on the finished panel.

Embedded masonry veneers consist of thin clay or concrete units with the appearance of brick, stone, or architectural block. These are sometimes called *thin brick* and *thin block*. The crew sets these facedown on the casting slab, in the desired pattern, with special parts that hold them in position and prevent concrete from bleeding through to their faces. The concrete is cast over the back, embedding them in place. After lifting panels, the crew removes any exposed parts and washes the masonry off to provide the appearance of a brick, block, or stone wall.

Connections

All structural connections to tilt-up panels are normally designed and specified by an engineer. They are decided and drawn before construction.

Connection of the panels to the footing may or may not be required. In some situations the weight of the walls and the backfill against them are found to be sufficient to hold them securely in place. When a footing connection is required, it is usually by means of dowels protruding from the footing. These may be welded to some form of weld plate on the bottom of the panels.

The panels may connect to the floor slab by means of the *closure strip*. When the floor slab is originally cast, a strip of one foot to three feet around the perimeter is omitted. After the wall panels are in place, that strip is also cast to extend

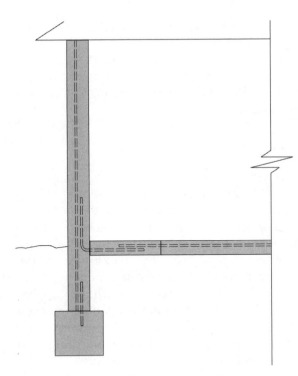

10-5 *Connections of walls to foundation and slab.*

the floor up to the walls. It is called the closure strip. The wall panels may have rebar extending out of them at the level of the floor. Likewise the original floor typically has rebar extending out of its edges into the closure strip area. When the closure strip is filled with concrete, it embeds the rebar extending from the panels and the floor, linking them together (Figure 10-5).

Most roofs in tilt-up construction are built on prefabricated steel joists. The joists connect to the top of the walls by way of weld plates or bolt plates embedded in the panels. Any floor decks installed inside the walls are typically also built on steel joists or steel beams. These attach to similar embedments in the wall by welding or bolting.

Many of the roofs in the far West of the U.S. are constructed with wooden trusses. These are usually attached to the walls by way of steel parts that are also welded or bolted to plates in the wall panels.

Architects

There are many architects familiar with tilt-up construction in the regions where it is common. These tend to be specialists in tilt-up design. Few architects with a general practice design an occasional tilt-up building. Some architects designing with tilt-up are members of the Tilt-up Concrete Association (TCA), and can be found on the membership directory on the organization's web site (www.tilt-up.org).

For buildings constructed in areas where tilt-up is not familiar, one common approach is to hire qualified architects from nearby regions. Other architects can and do learn the principles of tilt-up design and begin designing tilt-up buildings. The Tilt-up Concrete Association offers seminars to train architects. It also provides extensive manuals and technical briefs covering design. See the organization's web site for details.

Engineers

Tilt-up buildings are typically engineered in detail. Therefore, the engineer is important to a successful project. Panels must be designed to withstand lifting forces as well as the loads they will experience in the final building.

The manufacturers of tilt-up lifting hardware provide engineering services to insure that the panels and the hardware are appropriate to withstand the lifting forces.

The project's structural engineer is responsible for design to insure that the panels and the building can withstand the forces they will experience once they are in place. Most tilt-up design is done by engineers who do a large number of tilt-up buildings. Tilt-up engineering is not always part of a general structural engineering education, so not all structural engineers will be qualified. The engineers who do go to the effort to learn tilt-up engineering tend to make it a specialty and design a steady stream of tilt-up projects.

As with architects, one source for experienced tilt-up engineers is the membership directory of the TCA.

Training

Most tilt-up workers have learned the system through their work. To help newcomers, the Tilt-up Concrete Association now offers educational classes at its annual convention and at seminars it holds from time to time. These events are listed on www.tilt-up.org.

There are also independent organizations that provide training to contracting firms interested in beginning tilt-up construction. This training can be comprehensive. These organizations may also assist on-site in the firm's first project.

It is generally considered important to retain qualified and experienced tilt-up workers. This is particularly true for supervisors. Fortunately, there is now a certification program to help employers recognize qualified tilt-up workers. The program was developed by TCA and ACI. In it, a worker may become registered as an "ACI Certified Tilt-up Technician" or "ACI Certified Tilt-up Supervisor."

For the Technician certification, the worker must pass a comprehensive written test. The test is given a few times a year, usually at major concrete-related trade shows and events. To help the students prepare, there are usually optional preparation courses given earlier in the week at the same events. These preparatory courses are in addition to the many other TCA classes on tilt-up that may be offered.

To become a Certified Tilt-up Supervisor, the student must pass the test, and also show proof of a certain number of total hours of experience working on tilt-up construction projects. To maintain these certifications, the worker must re-take the test and present evidence of continued work experience every few years.

Further information on certification is available on the TCA web site (www.tilt-up.org) and the "Certification" page of the ACI web site (www.concrete.org).

Maintenance and repair

Tilt-up walls are highly durable and require little regular maintenance. The sealant in the seams between panels must be replaced every 10 years to 15 years, according to the manufacturer's specifications. When this is done, reports of any water intrusion in tilt-up buildings are rare.

If the panels are painted they must typically be repainted about every 5 years to 7 years to maintain their appearance. Stucco-type finishes may also require cleaning or repair periodically. Embedded masonry and exposed aggregate finishes typically require little if any maintenance.

Energy

Tilt-up wall panels have the energy-saving features typical in concrete walls: high thermal mass and airtightness. Their R-value can be anything from about two to thirty-two, depending on whether they are insulated and with what amount of insulation.

When a panel is insulated as a sandwich panel, most of the concrete is on the interior side of the insulation layer. This is because the structural concrete wall, which is typically 7 inches to 12 inches thick, is normally cast to be on the interior side of the panel. Engineering simulations indicate that putting the mass on the inside leads to the greatest reduction in the heating and cooling load from the thermal mass effect.

It is also possible to produce an insulation layer that is broken only by an occasional plastic connector. This requires some attention to detail at the panel joints, but it is not particularly difficult or expensive.

For these reasons, a well-insulated tilt-up panel is likely to be among the most energy-efficient of wall systems. The efficiency of the entire building is dependent on all building components including the roof, doors, and windows as well as the walls.

Sustainability

Tilt-up buildings should be able to qualify for the leadership in energy and environmental design (LEED) credits that are commonly open to concrete buildings.

If the building is well insulated, it will likely help qualify for multiple energy efficiency points.

The walls may help qualify for use of recycled materials if the concrete contains such items as fly ash or slag, or the reinforcement includes recycled steel.

The concrete almost always comes from the local area, and often the ingredients of the concrete do as well. This can be a major factor in qualifying the building for the local or regional materials credit.

Aesthetics

In the past, tilt-up buildings consisted of a set of concrete planes arranged in a "box." However, the designers of newer buildings are modifying the basic tilt-up panel and the way the panels are oriented.

Some contractors can now cast radius panels to create curved walls. Others adapt their formwork for nonrectangular panels. Panels are often connected to the building out-of-plane or out-of-vertical to add depth to the building or turn it into a virtual sculpture (Figure 10-6).

10-6 *Tilt-up building using extra panels as a decorative element.*
Chameleon Cast Wall System, LLC

Table 10-2. Finishes Available for Tilt-Up Buildings.

Finish	Description
Textured paint	Durable concrete paint with fillers that give the finish a substantial texture.
Colored concrete	Integral pigment in the concrete to provide almost any color permanently and without maintenance.
Exposed aggregate	Removal of the outer layer of cement paste to reveal the stones in the concrete, which may be selected for color and finish
Stucco-type finishes	Troweled-on finishes in almost any color. Can be textured and applied over foam moldings.
Form liners	Contoured and textured plastic surfaces set on the casting bed to give the concrete any of a wide range of finish surface shapes.
Thin brick	Half-inch thick clay brick embedded into the face of the concrete to create a finish of genuine brick.
Thin block	Two-inch thick block embedded into the face of the concrete to create a finish of genuine architectural block.

In recent years tilt-up has gained a range of optional finishes that extend far beyond plain concrete. Table 10-2 summarizes the major ones. The available finishes provide a wide range of options for tilt-up.

Key considerations

In many cases the greatest loads applied to tilt-up panels are during lifting. They must be designed with this in mind. The tilt-up accessory suppliers are helpful in evaluating this. They also provide panel layouts indicating the locations of the lift inserts and other information critical for erecting the panels.

Lifting the panels is a critical phase of construction. Careful planning is important to make sure it goes safely, efficiently, and without damaging any of the panels. This planning has to take into account such things as surface treatments, possible weather conditions, and crane availability.

With a small building that has panels formed by stack casting, the locations of the panels on the slab and in the stack are critical to the efficiency of the forming and lifting. Plan to spend considerable time working out the best arrangement.

When the panels are up, the temporary bracing is very important. The tilt-up accessory suppliers who provide them are also helpful in this. They do engineering calculations to determine the proper bracing, and provide the parts and diagrams of the bracing layout.

Replacing the sealant in the joints of the panels every 10 years to 15 years is important for preventing water leakage.

As with other concrete shells, tilt-up walls can be very air-tight and (when insulated) energy efficient. If the other components of the shell are similar, it is important to have an HVAC contractor or engineer familiar with air exchange and proper equipment sizing.

Availability

The materials needed for tilt-up construction are available anywhere.

The other critical item required is an appropriate crane. This will be a rubber tire or crawler crane with a sufficient capacity to lift all panels with a safety margin of 3 to 1. Such cranes are widely available, but when performing tilt-up for the first time in an area it is important to check on the availability of a crane of the right size and type for the job.

Crew availability

Tilt-up construction is spreading steadily across North America. In areas where it is practiced regularly, there are many general contractors who build with it and subcontractors who perform the work. However, not of all these are experienced with small buildings.

In areas without established tilt-up crews, or without any crews that are capable of producing a small building, there are some options. It may be possible to import a suitable crew, at some added expense. An existing crew may do additional research or do additional training. This will of course take time and add some cost. There are independent tilt-up training organizations that help a new crew or help an existing crew learn new skills. A good place to start locating crews or trainers is TCA.

Support

Since nearly all tilt-up buildings are fully engineered, the project engineer provides support on structural issues. The architect may do the same on design issues. The accessories company that provides the lift inserts and bracing typically has a skilled staff of engineers and other specialists that can help with many problems.

The Tilt-up Concrete Association offers detailed instruction manuals, as well as technical briefs on particular aspects of tilt-up construction. They are listed on www.tilt-up.org.

Current projects

The most likely option for locating nearby tilt-up projects is asking locals in the business. Tilt-up contractors and the local representatives of the tilt-up accessory companies are frequently aware of current projects. The members' directory of the TCA includes many contractors. It also includes the accessory companies, who can direct you to their local distributors.

11

Autoclaved, aerated concrete walls

Autoclaved, aerated concrete (AAC) is full of tiny air bubbles. Produced in a factory, about half the volume is air. The rest is a concrete with fine sand aggregate and no stone. The air bubbles make AAC relatively lightweight and a good insulator.

Autoclaved, aerated concrete is produced in two types of construction units, blocks and panels. The material is cast as a large billet, and then wire-cut into these shapes. The blocks can be assembled into walls with methods similar to those used for concrete block. The panels assemble much like grouted precast wall panels (Figure 11-1).

The methods for producing AAC were developed early in the twentieth century. It has been used for walls for decades in Europe. However, the products have only been available in North America for about ten years.

Autoclaved, aerated concrete walls combine many functions in one material. They act as insulation and part of the wall structure. The surface of the material can be routed to create chases for electrical cable. Finishes may be troweled directly to the surface. The result is a wall that consists almost completely of concrete all the way through. Nearly the whole wall shares concrete's properties of durability and stability as a result. Autoclaved, aerated concrete can also be cut more readily than conventional concrete and will accept nails and screws more readily. Since it is relatively light and easy to shape, AAC can also be practical for constructing interior walls.

Autoclaved, aerated concrete use in North America has gone from zero in 1993 to about 6 million square feet per year currently. Most of this is for use in building walls. Currently AAC is used primarily for commercial building, mostly in the Southern United States. However, it is expanding to the residential market and more northern areas as well.

11-1 *Wall and floor components made of autoclaved aerated concrete.*

History

A Swedish scientist discovered the process of creating cellular concrete in 1923. He found that adding aluminum powder to cement, lime, water, and finely ground sand filled the mixture with air bubbles, causing it to expand dramatically. This made for a product that had some structural strength but was still lightweight. It could also be produced with fewer raw materials than the same volume of conventional concrete. Products using this technology first became widely available in Germany after World War II. It is now common throughout much of Europe, but the American market was first introduced to the product much later.

Based on the success in Europe, investors constructed new plants in the U.S. to tap the North American market. The three plants currently operating are located in the Southeast and the Southwest, and most sales are in these areas.

In the late 1990s, the producers formed the Autoclaved Aerated Concrete Product Association (AACPA). This trade association has worked to publicize AAC, get it included in building standards and codes, and promote research and the development of related products.

Market

Current sales volume of AAC wall materials is estimated at several million square feet per year. About 75 percent of this goes to commercial and industrial construction and about 25 percent to residential. The commercial and industrial buildings are nearly all low-rise. The greatest use is in hotels and motels, schools, libraries, and some warehouse and manufacturing facilities. Most residential use is in high-end, single-family homes.

Sales are primarily within a few hundred miles of the existing manufacturing plants. Currently one plant is in Arizona and two are in Florida. Material has been shipped across the United States and to points in Canada.

Advantages to the owner

Autoclaved, aerated concrete walls have most of the benefits of concrete walls. In some cases they have them to an even greater extent than conventional concrete.

The sound reduction of AAC walls is high. It allows less sound through than some other walls that are thinner, such as frame walls.

The combination of insulating properties and concrete material also give AAC especially high fire resistance. In the standard firewall test, eight-inch AAC receives a four-hour rating, which is the maximum assigned to a wall assembly. It actually meets the test requirements for up to eight hours. It is also a highly noncombustible material. If fire penetrates the wall finishes, the AAC will not burn or give off significant fumes.

Since AAC is relatively easily cut and shaped, building design is fairly flexible. Unusual shapes and surface features can be created relatively inexpensively by cutting or shaving off material

The walls provide good energy efficiency. Their R-value is comparable to that of a conventional frame wall, but their thermal mass and air tightness are higher.

Autoclaved, aerated concrete needs a good, protective finish material since it is softer than conventional concrete. However, it resists water, rot, mold, mildew, and insects like other cementitious materials.

It has a lower strength than conventional concrete. However, the required building strength can be attained by creating reinforced concrete columns and beams inside it with grout and rebar.

Advantages for the contractor

The light weight of AAC makes it easier to handle than some other building materials. The block products are fairly easily learned by masons, and the panels can generally be handled by conventional precast crews using the same methods.

Components

Standard AAC blocks are usually 8 in. high and 24 in. long. They come in thicknesses of 4 in., 6 in., 8 in., 10 in., and 12 in. Unlike conventional concrete block, they have no cavities. A standard 8 in. × 8 in. × 24 in. AAC block weighs about 33 lb.

The properties of AAC as it is manufactured in the United States are approximately as listed in Table 11-1.

Special blocks have a U-shaped cut-away along the top. They are used to form bond beams or grouted lintels. They come in thicknesses of 8 in., 10 in., and 12 in.

Some manufacturers also offer *tongue and groove blocks*. They are used as conventional blocks. However, they do not require mortar at the end joints since they overlap there. They come in the same sizes as the standard blocks.

There are also *cored blocks* to create reinforced grouted cells. These blocks have cylindrical vertical cavities precut in them.

Wall panels are generally designed to be installed vertically. Autoclaved, aerated concrete wall panels are a standard 24 in. wide by a height of up to 20 ft. They come in thicknesses of 6 in., 8 in., 10 in., and 12 in. They can have a concave depression along the two long edges (the sides). The depression is in the shape of a half-cylinder, so when two panels are set side-by-side, a cylindrical cavity is formed between them.

Four-inch thick panels are also available for use as interior non-load bearing walls.

A special thin-bed adhesive mortar is used to join the blocks to one another. It is thinner than conventional mortar and serves to "glue" the units together. It is applied with a special trowel that holds a large amount of the material and has teeth to spread a precise amount over the block surface.

Table 11-1. Properties of AAC.

Property	Approximate Value
Fire rating, 4-inch minimum block wall	4 hours
Sound transmission class, 4-inch wall	40
8-inch wall	45
Density	25–50 pounds per cubic foot
Compressive strength	300–900 pounds per square inch
Allowable shear stress	8–22 psi
Thermal Resistance (R-value)	0.8–1.25 per inch thickness

The wall assembly

Autoclaved, aerated concrete walls include the AAC material itself, and the reinforced grouted cells and beams running through it. The configuration of the AAC elements is different for blocks versus panels. The configuration of the grouted elements is virtually identical regardless of which product is used.

Autoclaved, aerated concrete blocks are stacked in running bond, much like conventional concrete blocks. They may be cut as needed to achieve the designed wall lengths and heights. They are also stacked up to openings like conventional blocks. An opening may have a buck. The buck will typically carry the blocks above it. It may also have no buck. In this case it will usually have an AAC lintel above, spanning the opening. The blocks are connected to one another with a thin layer of the special AAC mortar.

Autoclaved, aerated concrete panels are usually stood vertically. They are sized to extend from the floor to the top of the wall.

Regardless of whether the wall is constructed of blocks or panels, it contains reinforced grouted cells and beams. The positions of these are determined by the structural strength the walls require. Typically there is one vertical grouted cell at each corner, one along either side of an opening, and one in the field of the wall about every six to eight feet. There is one bond beam completely around the top of the walls. There may also be a shorter beam over an opening, connecting to the grouted cells on either side. Each cell consists of conventional grout with one steel reinforcing bar in the center. Beams consist of conventional concrete with usually two reinforcing bars, one above the other.

Electrical cable and boxes are let into the AAC material. The cable runs in chases cut in the surface. The boxes fit more or less tightly into rectangular cutouts on the surface. Building codes generally require that the cable be in conduit, instead of in contact with concrete. In this case the conduit is installed in the chase and the cable pulled through later (Figure 11-2).

Nonstructural walls, such as interior non-load bearing walls are pure AAC. They do not typically have any columns or beams inside. They are attached with the adhesive mortar along their edges to the foundation or floor below and to the other panels alongside.

Cost

Subcontractors charge about $10 to $12 per square foot for structural AAC walls. In small buildings it will be somewhat more than wood frame walls.

A typical breakdown of the subcontractor's costs (not including his markup) is in Table 11-2.

The total subcontractor installed cost for AAC block walls ranges from $10 to $12 per square foot, including the subcontractor's markup. The pricing varies greatly based on the type of wall, the complexity of the wall and the size of the project.

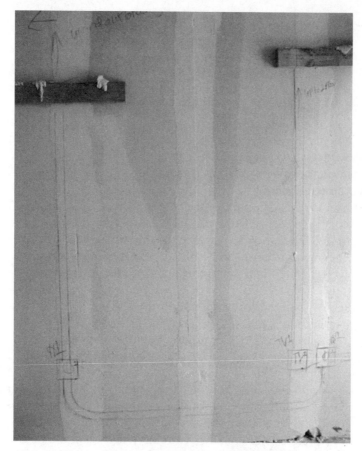

11-2 *Walls marked for chases (top), and electrical conduit and boxes installed in chases cut in the AAC (bottom).*

Table 11-2. Approximate Costs of Autoclaved, Aerated Concrete Wall Construction.

Block Walls

Item		Cost per Square Foot*
Labor		$3.00
Materials		$4.30
AAC block	$3.00	
Rebar	$0.35	
Grout	$0.15	
Miscellaneous	$0.80	
Pump rental		$0.25
Total		$7.55

Panel Walls

Item		Cost per Square Foot*
Labor		$2.00
Materials		$6.50
Panels	$5.75	
Rebar	$0.35	
Grout	$0.15	
Miscellaneous	$0.25	
Pump rental		$0.25
Crane rental		$0.35
Total		$9.10

*All costs in U.S. dollars per gross square foot wall area.

The cost of block construction is frequently more in large buildings. Work on taller walls requires more extensive scaffolding and is slower. The cost of panel construction is frequently lower for larger buildings and building with repeated panels. In these applications the installation of panels can be repetitive and more efficient.

Costs also vary by region. The farther the project is from an AAC plant, the greater the shipping costs that will be added to the material price. The current manufacturers are in Florida and Arizona.

Code and regulatory

Autoclaved, aerated concrete construction is not covered in the International Building Code, International Residential Code, or National Building Code of Canada. Almost all AAC buildings constructed in North America are engineered, and the designing engineer supplies full documentation to the building department. The manufacturers can supply extensive documentation. This includes complete evaluation reports from the International Code Council (ICC).

The Autoclaved Aerated Concrete Product Association (AACPA) is working with the American Concrete Institute (ACI) to develop building code guidelines for construction with AAC. The 2005 edition of ACI 530 includes an Appendix specifically for AAC masonry. American Concrete Institute Subcommittee 523A is developing a document entitled "Guide for Using Autoclaved Aerated Concrete Panels" to be used in conjunction with ACI 318. The next step will be to have it adopted as part of the major building codes, like the rest of ACI 318.

Installation

Construction of AAC walls differs depending on whether blocks or panels are used, and whether the walls are load bearing or not.

Load bearing block walls

Block walls are installed directly on the foundation, or on an elevated floor in multistory construction. The workers lay conventional mortar on the foundation or floor for the bottom of the first course. It is laid thick enough to level the course precisely. On all work above this initial leveling bed, the mortar used is the special AAC thin-bed adhesive mortar. The ends of the blocks are bonded with the mortar, unless tongue and groove blocks are used. Since most of the blocks have a solid top face, the mortar should cover the entire surface. The blocks are laid in a running bond with at least 6-inch overlap, also similar to standard block masonry (Figure 11-3).

11-3 *Mason installing AAC Blocks.* AERCON Florida, LLC

In the locations where vertical grouted cells are to go, the masons must install cored blocks. If necessary, they can make them from solid blocks by cutting or drilling out the hole. Typically there will be dowels extending out of the foundation or floor at these locations. The cores thread over these dowels.

The masons adjust the blocks to plumb, level, and square with a rubber mallet while the mortar is still pliable. Where necessary to accommodate odd dimensions or angles, the blocks are cut with a handsaw or band saw. The upper courses are set in much the same way. At grouted cell locations it is important to align the vertical grouted cells of blocks on different courses precisely.

At openings, blocks are cut as needed to fit up to the jambs on each course. The blocks are cored to create vertical columns along either side of the openings. A pre-manufactured lintel goes over the top of the opening, spanning its width plus an additional minimum bearing length on either side that is specified by the engineering.

Alternatively, it is possible to install U-blocks over an opening. In this case there must normally be temporary shoring installed to support the blocks above. U-blocks also go on the top course of the wall.

Vertical rebar are placed in the vertical cells and lap the dowels below. Horizontal bars go in the U-blocks over the openings, and in the bond beam around the top of the walls.

When all blocks and rebar are in place, the crew wets the cavities. This is necessary to prevent the blocks from drawing large amounts of moisture from the concrete and degrading the mix. They then pump grout into all the vertical and horizontal cavities and consolidate with a vibrator.

Non-load bearing block walls

Interior walls are often built with a narrower block, usually 4 in. or 6 in. wide. They are constructed in the same way as load-bearing walls, but without any cavities, rebar, or grouting. These walls may be attached to the ceiling or roof members above with metal straps.

Load bearing vertical wall panels

The installation of load bearing vertical wall panels typically starts at a corner. The panels are lifted into place with a crane, one at a time (Figure 11-4).

The panels are set on a bed of the thin-bed mortar laid on the foundation or floor below. Extending from the foundation/footing are dowels. The panels have concave edges at the dowel locations. The dowels are positioned and the panels sized so the concave edges fit around the dowels. At this point the crew ties a vertical bar to the dowel. When the neighboring panel is installed, the concave edges on the two adjacent panels form a hollow cylindrical column around the dowel and the vertical bar.

11-4 *Installation of AAC vertical panels.*

A small 1 in. × 1 in. corrugated sheet metal piece is inserted near the top and bottom of each two adjacent panels to lock them together while the thin-bed mortar is setting.

Window openings may be pre-cut in the panels, or the crew may cut them out on-site.

The bond beam on top of the wall may be created with plywood and 2-inch thick flat AAC material, or with bond beam blocks. Typically the flat material is used when an AAC deck goes on top, as discussed in Chapter 18. Bond beam blocks are generally used when a frame roof goes on top (Figure 11-5).

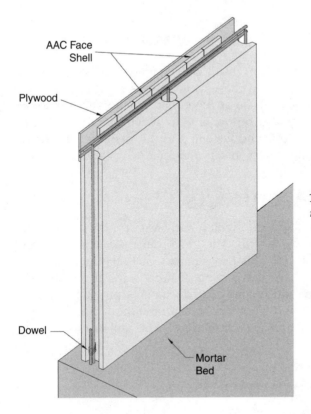

AAC Face Shell

Plywood

Dowel

Mortar Bed

11-5 *Wall of AAC panels ready for grouting.*

Non-load bearing vertical interior panels

Vertical wall panels are sometimes used to create interior walls that are non-structural. Light panels may be manually placed, while larger ones are craned. They may be shimmed to precise height and plumb with mortar packed below, or set directly in a mortar bed. A steel rail or angle is fitted along the top of the panels for lateral support.

Connections

A frame roof may be connected by either the conventional top plate or hurricane strap method. The connectors are embedded in the bond beam along the top of the wall. A heavy steel roof may be attached to weld plates or bolt plates embedded in the bond beam.

Frame floors may be attached with the standard ledger method. Heavy steel floors may be attached to weld plates or bolt plates. In either of these situations, the floor level would have to align with a bond beam so that the anchor bolts or the plates could be embedded in the beam. However, many AAC projects also use AAC floors. In this case the construction resembles platform framing. The lower-story walls are created, precast floor planks are set on top of them, the upper-story walls are built on top of this, and so on. The details are in Chapter 18.

Light fixtures may be attached to AAC walls with common wallboard screws. Care must be taken not to overtighten these, however, as they may strip out the AAC material. Heavy fixtures can be attached with anchors into the concrete columns or beams. It may also be possible to attach them by adhering plywood to the surface of the AAC and fastening to the plywood. There are also heavy-duty anchors manufactured specifically for AAC.

Architects

Normally architects adapt readily to designing with AAC products. They are similar to other systems they work with. The block dimensions are different from those of conventional blocks. This may require some getting used to.

The ease of cutting and shaping the product adds some flexibility. The evaluation reports and design and engineering documents provide architects with a substantial amount of information for such a new product.

Like any new product, an architect's first use of AAC will require time to learn its unique characteristics.

Engineers

Engineers also usually adapt easily. The product has such strong similarities to precast and block construction that the step is not a big one to make. The engineering is similar to conventional precast and masonry engineering.

The manufacturers have complete engineering design guides and some of them have free engineering software available through their web sites. The AACPA also has engineering software available.

Currently, structural design for block construction can be done according to ACI 530. The 2005 Edition contains an Appendix specifically for AAC masonry design. In the future AAC is expected to have its own section of ACI 318 and that may be adopted by the building codes. Manufacturers have engineering staff that can help engineers using the product.

Training

Usually one or two knowledgeable people are enough to lead a small scale commercial project and most residential projects. Manufacturers offer training through seminars or by sending people to the job site. The easiest way to access registration is through their web sites. The manufacturers can be reached through the Autoclaved, Aerated Concrete Products Association (www. aacpa.org).

Because of the product similarities, experienced masons and precast installers adapt most easily to AAC. One day of training may be enough to learn the essentials of the new system. However, classroom training cannot prepare an installer for every detail that comes up at the job site. It is advisable for a first-time crew to have an experienced person assisting on at least their first project. The manufacturers frequently provide this kind of help.

Maintenance and repair

There is no routine maintenance required for a properly finished AAC wall. Like most concrete walls, one built of AAC is stable. It is resistant to shifting or sagging over time, and to insects, vermin, or changes in temperature.

The material must ordinarily be finished on the outside, however. It is porous and will retain moisture, so it should not be left exposed. If a stucco-type finish is desired, the manufacturers recommend certain products formulated specifically to go over AAC. These materials contain some acrylic polymers in the mix for good sealing against liquid water, but they also breathe enough to allow water vapor to escape. Almost any conventional siding product or masonry veneer can be used as well.

Connections made into the AAC must also be made with some care. They could be detrimental if they open up holes for water to enter. In addition, since the AAC material is highly stable, it will not expand or contract much as the temperature changes. Similar to any building material, if fasteners installed into it do expand and contract, they might shift or work loose over time.

The AAC surface might be dented by a hard impact. If so, the dent can be filled with AAC patch products and refinished.

Energy

Tests of actual houses with AAC walls indicate annual heating and cooling savings in temperate climates of roughly 10 percent to 20 percent. This is compared to matching houses built with conventional wood frame construction. Test houses in northern climates indicated similar savings during the warm months, but somewhat lesser savings in the winter. This is as one would expect, because the savings from thermal mass are less during consistently cold periods. However, engineering studies suggest that there will still be total-year savings even in the coldest climates of the United States.

The R-value of AAC is approximately 1 per inch. This results in an R-value of about 8 for an eight-inch wall. There are no significant breaks in the insulation. The R-value will be reduced somewhat where there are grouted columns or beams in the wall.

The thermal mass of the AAC wall is fairly high, and it is well exposed to the interior of the building. The walls have generally been found to be very airtight.

Sustainability

Autoclaved, aerated concrete wall construction can increase the energy efficiency of a building. Although this is not usual in North America, the material can be manufactured partially with fly ash, which is a recycled material. AAC also has an environmental advantage of containing less total raw material than conventional concrete.

On the Leadership in Energy and Environmental Design (LEED) scale, AAC walls can help a building qualify for one or more points awarded for energy efficiency. The total points received on this scale will depend on other energy-related parts of the building as well.

Any use of fly ash in the AAC and use of recycled steel in the rebar may help qualify for the point awarded for recycled materials.

Many buildings currently constructed with AAC could also qualify for the point for use of local materials. Most projects are within the 500-mile radius allowed by the LEED scale. Projects far from the Southeast or Southwest might not qualify.

Aesthetics

Irregular shapes such as curves, non-ninety-degree corners and the like are relatively easy to create with AAC because it is easy to cut and shape.

Any conventional finish is feasible on AAC. Stucco-type products are most common because they install quickly and inexpensively. No water-resistant layer or mesh need be installed. The troweled-on finishes adhere well to the AAC.

Brick or other masonry veneers may be added to create a cavity wall. Standard installation applies. The brick ties can be fastened to the AAC.

Any sidings attached with nails or screws can also be installed on AAC. To do this, the crew first fastens furring strips to the wall with drywall screws. They then fasten the siding to the strips. This is less common because of the added expense of the furring.

Key considerations

In building block walls, alignment is critical. The cavities for grouted cells must be cut correctly and must line up all the way up the wall.

On panel projects, keeping close track of the panels is important. The factory provides detailed shop drawings showing the correct location and placement of each panel. If any panels are interchanged, the correct panels will show up missing at some point.

As with other energy-efficient concrete wall systems, heating, ventilation, and air conditioning (HVAC) contractors may tend to oversize the equipment for an AAC building. The dangers of this are unnecessary expense and equipment that short cycles. The remedy is to retain a contractor experienced with energy-efficient construction or sizing software, obtain the software available from the Portland Cement Association (www.concretehomes.com) or retain a mechanical engineer (as is typical on commercial projects).

The tight walls may also bring the air infiltration down to a low level. This may call for an HVAC system that introduces a controlled amount of fresh air. It should be discussed with the HVAC contractor or mechanical engineer in charge.

Availability

Three manufacturers produce AAC in the United States. Their plants are located in Arizona and Florida. They will ship anywhere, but costs increase with distance. For that reason, most projects that consider AAC are in the South. Some projects may also import material from Mexico or Europe in cases where that is less expensive.

The Autoclaved Aerated Concrete Products Association (AACPA) has a list of distributors around the country. They can be reached through their web site, www.aacpa.org.

Crew availability

There are some crews experienced with AAC, mostly in the South. The product manufacturers and distributors can provide contact information for them.

Many projects proceed with a newly trained crew. The leader of the crew should receive training from the AAC supplier. There should also be an experienced person on site to help with the first job. The supplier can usually supply this person.

The new crews most often used for AAC block jobs are masonry crews, and for panel jobs they are conventional precast crews. Masons have sometimes complained of difficulties adapting to AAC block work. They cite the larger block size, the lack of regular cavities for ease with gripping the block, and the different mortar, trowel, and leveling technique. Carpentry crews have successfully adapted in some cases, and provide another option.

Support

If your manufacturer has a local distributor, that is usually the best initial contact person for support. The distributor is usually on call and frequently comes to the job site for first time contractors. If there is no local distributor, call the manufacturer directly. The manufacturers have manuals, evaluation reports, and engineers on staff to answer questions.

Current projects

The best method for locating current AAC jobs to visit is to contact the US manufacturers or their local distributors. They can be contacted through the AACPA web site, www.aacpa.org.

12

Developments in concrete walls

There are many other developments in concrete walls that could be useful to certain contractors or in particular projects. Some of these are other wall systems that are new or offer unique capabilities. Others are novel methods or materials that offer some significant advantages in at least some cases.

Mortarless block

There are newly designed concrete block systems that stack up without mortar. The blocks are sized and shaped so that they sit directly on top of one another and interlock into position. They are then fixed together by various means, such as by grouting, post-tension rods, or an adhesive applied to the joints or surface.

There are several potential attractions of mortarless block. Without the trowel work and precise hand leveling of conventional masonry, work can go faster and with a lower labor expense. The workers can usually also be less skilled, further reducing costs. The installation need not be as often delayed or altered for bad weather, because there is no mortar to get wet or freeze. The cost of the mortar itself and mixing and handling it are also eliminated (Figure 12-1).

In other respects the wall produced with mortarless block can be much like conventional block walls. Most systems permit grouting and reinforcement, to bring the wall to the level of strength desired. The mortarless units are typically in 8-inch dimensions. Some systems come with special provisions for insulating, or conventional masonry insulation methods can be used.

Mortarless block is less familiar than conventional block to much of the market. However, some of the systems are supported by extensive engineering and evaluation reports.

12-1 *Stacking mortarless blocks.* Azar Mortarless Building Systems, Inc.

There are manufacturers of mortarless blocks across North America. For more information, contact the National Concrete Masonry Association (www.ncma.org) or Masonry Canada (www.masonrycanada.ca).

Foam-faced tilt-up

Promising new developments for the use of tilt-up in small buildings are foam-facing systems. They can make construction less expensive and more efficient by combining several steps of wall construction.

The construction process is similar to conventional tilt-up. However, the bottom of the casting bed is lined with a layer of foam. The foam may be out-fitted with steel or plastic inserts that will serve as studding for later attachments. It may also be contoured so that concrete is cast only where it is needed.

After these parts are in position, assembly is more or less conventional. Workers install rebar and lift inserts and cast the concrete. The panels are lifted into position. However, their bottom sides are turned toward the interior of the building. This puts the insulation and studding toward the interior and the durable concrete face outside (Figure 12-2).

Because the panels are cast over a layer of foam, no bond breaker is required and the need for a blemish-free casting surface is eliminated. Stack casting can proceed with fewer steps. Contouring the foam allows foam-faced tilt-up to reduce the amount of concrete and reinforcing steel by up to 40 percent compared with the conventional panels. This reduces materials cost. It also allows the use of a smaller, less expensive crane and makes the lift go faster.

When the panels are up, the walls are already insulated and studded. These are two features commonly required on small buildings. With these systems they require no additional steps or work high off the ground. Chases for electrical lines may be cut into the foam. Wallboard or other interior finishes can be attached. Insulation values have no upper limit. Panels have been constructed with foam thicknesses that provide an R-value of up to 30.

The erected wall panels provide a solid concrete substrate for the application of various exterior finishes, such as spray-on textured finishes and stucco-type products. Many traditional finishes may also be applied.

A number of buildings have already been constructed with foam-faced tilt-up systems and more are on the way. The developers are working to enhance the architectural finish options and widen the application of the product to new building types.

12-2 *Lifting a foam-faced tilt-up panel.* Avantec, LTD

Vinyl stay-in-place forms

A useful product that is becoming well established is the rigid concrete form made of vinyl plastic that stays in place permanently. It offers efficient construction of a high-strength wall with a durable built-in finish on both the interior and exterior face.

The systems generally consist of long, hollow vinyl members that are slid together side-by-side to construct a form wall. They are braced in position and can be outfitted with rebar much like a conventional concrete wall. The concrete is then placed vertically through the top of the wall panels (Figure 12-3).

12-3 *Assembling vinyl stay-in-place formwork.* Royal Building Technologies

Some systems offer a pre-installed insulation cavity on one side of the forms for an R-value of up to about 22. Thermal mass and airtightness are high. Also available are raceways that can be pre-installed for easy insertion of electrical wiring at a later time.

The finished wall assembly has a low maintenance layer of vinyl on each face. This can remain as the permanent finish, or conventional finishes can be attached. The vinyl is highly resistant to dirt and the elements and is easily washed.

One popular application of the system has been agricultural buildings, which need to have an uncontaminated surface that can be washed down periodically. Another has been car washes, which must endure repeated exposure to water and detergents without deteriorating. However, the system has been used for a wide range of residential, commercial, and industrial buildings.

Shotcrete walls

Another technology developed in Europe that is being adapted to North America is the *shotcrete wall panel*. Shotcrete is the process of shooting wet concrete onto a wire or rebar cage to create a surface. The concrete collects on the steel and workers trowel it smooth to produce the desired shape and texture. The result is reinforced concrete in any shape the trowel can be moved. A variation on this process is already used extensively in North America to produce swimming pools.

The modern shotcrete panel system starts with a layer of foam. A factory puts a layer of welded wire mesh on either side of the foam. Special wire crosspieces pierce the foam to connect the mesh layers. The final unfinished panel consists of a layer of foam with a layer of mesh about an inch away from the foam on either side, and a rigid steel connection between the mesh layers.

Crews erect these panels into the shape of the exterior walls of a building. By cutting and bending, they can contour the panels into unusual and interesting surfaces. A layer of shotcrete about $1^{1}/_{2}$ in. thick is applied to each side and troweled smooth. The resulting walls consist of two thin layers of closely reinforced concrete that sandwich a layer of insulation foam. The concrete provides a hard shell on either side. The insulation can be adjusted in type and thickness to produce any of a wide range of R-values (Figure 12-4).

Shotcrete panels are available now from U.S. manufacturers. They are particularly used in applications that need an insulated wall with either a hard shell on both sides or free form shaping. The system is thoroughly tested, and the manufacturers have evaluation reports from major code bodies. There are now crews trained in the construction.

12-4 *Applying shotcrete to walls formed with shotcrete panels.* Hadrian Tridi-Systems

Aerated concrete sheathing

Some manufacturers of aerated concrete offer 2-inch thick panels to use as sheathing over frame buildings. This offers a high-quality backing for finishes that is durable and has insulating value.

The panels are typically a foot or more wide and several feet long. When a wood or steel building frame is in place, the panels are set horizontally around the outside of the framing. They can be mechanically attached to the frame (Figure 12-5).

The panels replace sheathings of foam, plywood, oriented structure board or faced gypsum boards. They have an attractive combination of features that none of these other products exactly matches. The aerated concrete material

12-5 *Building with aerated concrete sheathing.* AERCON Florida, LLC

is resistant to water, fire, rot, rust, mold, mildew, and insects. The layer adds an R-value of about 2 to the building envelope. It also adds thermal mass to the shell, further improving the building's energy efficiency. The material is easily tooled and shaped. As a result it is easy to add contours and relief to the surface, such as rounded corners, reveals, and decorative inscriptions. Extra material can be adhered to the surface to create positive relief such as decorative moldings. Although not as hard as standard concrete, the aerated material offers more resistance than some other sheathings.

The aerated concrete is readily finished with troweled-on products, generally without the addition of a weather-resistant membrane or mesh. Other finishes can be attached by various means.

Aerated concrete sheathing is now well established and practiced, especially in the South and Southwest of the United States. For more information, contact the Autoclaved Aerated Concrete Products Association (www.aacpa.org).

Safe rooms

A safe room is a way to get protection from high winds in a home or small building constructed of lightweight materials. It is a single, wind-resistant room that gives occupants of a building a place to ride out a hurricane or tornado—even if the rest of the building is not particularly wind-resistant. Although it is possible to build safe rooms out of different high-strength materials, concrete is one of the most practical options.

A walk-in closet is a popular choice for the safe room. It is large enough to hold a few people, has no window and usually only one door.

Safe rooms are nearly always constructed on a slab foundation. Steel rebar dowels in the slab extend into the safe room walls to make a high-strength tie-down for the safe room. The ceiling is also concrete, formed with one of the deck systems described in Part 3. The wall rebar extends up into the ceiling to lock everything together. A heavy-duty steel door completes the package.

The U.S. Federal Emergency Management Agency (FEMA) publishes a guide to the construction of safe rooms that includes plans for building them out of different concrete wall systems. It is called, "Taking Shelter from the Storm: Building a Safe Room Inside Your House." It can be downloaded for free from the FEMA web site (www.fema.gov) or ordered in hard copy version from the same site.

Self-consolidating concrete

Self-consolidating concrete (SCC) promises some significant potential improvement in the speed and cost of many concrete wall systems.

The difference will be hidden from the contractor's view in the case of precast panels. However, SCC can help precasters fill their forms faster, with fewer workers involved in moving the concrete into position.

For any system with concrete placed on-site, the contractor's labor will decline and quality assurance will go up. When producing grouted masonry, ICF, removable forms, or tilt-up walls, the forms will fill up without the workers having to move the concrete hose. The tilt-up forms will fill with few if any workers raking the concrete into place. For the vertical systems, it will be unnecessary to have a worker on a scaffold moving to fill the wall at many different points. A hose can be hooked up to the wall at a few points, and the concrete can run in like honey filling a jar. In all of these cases, no one will need to run a vibrator to consolidate. Yet the concrete should be virtually void-free.

It will take some time for SCC to reach its full potential. The ready-mix suppliers must learn the fine points of these precisely controlled mixes before they can supply them reliably. The crews need to learn how to control the concrete and the rate of pour to be efficient without overstressing or overfilling the forms. However, over time SCC is likely to have a positive impact on the cost, practicality, and quality of concrete walls.

Structural fiber reinforcement

Producers are exploring the use of synthetic fibers as structural reinforcement in concrete walls. This could change the way walls are constructed in the factory and in the field.

The attraction is high. High-strength fibers mixed into the concrete may allow much of the steel reinforcing bar to be omitted from walls. This would eliminate buying, storing, cutting, bending, and installing the bars. It would also eliminate the problems of moving concrete around bars that obstruct easy flow into the forms.

Several small buildings have been cast at the job site with fiber reinforcement. The suppliers have performed significant tests and engineering on the application. At this point not all of the rebar has been eliminated. Some must remain, particularly at major stress points. Crews must get a feel for the flow and placement of the new mix, although several have reported that their pours went fine. Over time these things will become refined, and workers and designers are likely to become increasingly comfortable with the new material.

PART III

FLOOR AND ROOF SYSTEMS

13

Background on concrete floor and roof systems

Following the growth in concrete wall systems, there has been a growth in the use of concrete floors and roofs for small buildings. Floors and roofs are sometimes referred to as *decks,* and the systems used to construct them are called *deck systems.* The popular deck systems are diverse. Yet they also share some key properties with one another. Most of these common properties have to do with the nature of concrete and steel.

The deck systems that are currently most practical in small buildings include cast-in-place and precast methods. The precast deck systems use either conventional concrete or autoclaved, aerated concrete (ACC). They are similar, except for the material used.

History

Cast-in-place decks using modern concrete appeared in the late 1800s. They were created with wooden formwork. However, spans were limited until the art of steel reinforcing was better developed.

Concrete and steel decks have long been used in large buildings. For large buildings the rigidity of concrete decks helps create a structure strong enough to stand up to the severe forces acting on it. The decks tie the walls and interior columns of the building together so that they support one another. This eliminates the need to make each wall strong enough to withstand all forces on its own.

However, practical small buildings do not depend on having such rigid decks. These smaller structures could be made strong enough with lighter decks.

Therefore, for cost reasons the vast majority of small buildings have been built with light frame floors and roofs.

After the growth in the popularity of concrete walls for small buildings, concrete floor and roof systems better suited to small projects began to emerge as well. They come from many sources.

The new wave of concrete decks is a clear outgrowth of the new interest in concrete walls. Most of the customers that specify concrete walls for their small buildings have done so because they see specific advantages in the concrete material over the lighter alternatives. When they consider their floors and roofs, they want to get these desirable properties there as well. As concrete decks suitable for small installations are becoming more widely available, a growing percentage of buyers are choosing them.

Today about 5–10 percent of all small buildings constructed with concrete walls are built with concrete floors as well. Concrete roofs are somewhat less common. However, the use of concrete roofs has been growing significantly since 1990, when virtually none were built.

Market

The larger the building, the greater the chances that some of its decks will be constructed with concrete. Part of the reason is simple economics. The cost per square foot of constructing light-frame decks tends to go up as they get larger. In contrast, the cost of reinforced concrete decks may actually go down.

In addition, larger buildings usually require higher-strength decks. They tend to be for high-traffic and heavy applications.

Nonetheless, even among single-family houses the use of concrete decks is growing. In these very small buildings concrete is currently concentrated in the higher end of the market. Some of the most efficient deck systems are just starting to appear in the middle market.

Advantages to the owner

All concrete decks, both roofs and floors, have the durability advantages of concrete. They are nearly impervious to rust, rot, and insects. They tend to survive fire well. They resist sagging and shifting.

Roofs

For the building owner, the advantages of concrete roofs are virtually the same as the advantages of concrete walls. The walls and roof make up the total building shell. They separate the indoors from the outdoors, and the benefits that come from building with concrete in the walls are increased by extending the concrete to the roof. Buildings with concrete roofs are believed to be significantly more energy efficient, disaster resistant, quiet, and comfortable than buildings with only concrete walls.

A concrete roof can be particularly important in making a building resistant to wind damage. In high winds such as hurricanes, light frame roofs can detach from the rest of the building. Once the roof is gone, the walls lose an important source of support. Without a roof to connect them, light frame walls may collapse. Reinforced concrete walls typically remain intact. However, even in a building with concrete walls, when the roof is gone, there can be extensive damage to the interior, and the occupants are threatened. If the building had a reinforced concrete roof atop reinforced concrete walls, the potential for damage is sharply reduced.

Similarly, concrete walls typically transfer only about one-third as much sound energy as a typical wood-frame wall. If the walls are concrete and the roof wood frame, noises originating alongside the building will be muffled, but may still penetrate the roof more than the walls. When the roof is also concrete, the amount of sound entering is lowered another notch.

The increase in comfort comes about for the same reasons the walls increase comfort. A concrete roof allows little air through, it will (in some systems) be well insulated, and it provides substantial thermal mass to the shell that helps to smooth out the temperature fluctuations from outdoors.

Floors

In the case of floors, the advantages to the building owner are different. Floors do not separate indoors from outdoors. They do, however, separate different parts of the house. This may be desirable. For example, sounds produced on one story will be better contained, so that people on other stories are less disturbed. This is particularly important in buildings housing people who require quiet or concentration, such as hotels and motels, homes, and offices. Certain floor systems are also well insulated, so they help contain heat and cold. This may be important in buildings with zoned heating or cooling.

One of the greatest advantages of concrete floors is their rigidity. They flex and bounce considerably less that light-frame floors. This generally provides a more reassuring feel to the occupants.

A special advantage of concrete floors is how suitable they are for *in-floor radiant heating*. Radiant heating consists of heated fluid that runs through tubing around the building. The heat of the fluid radiates into the surrounding space, warming the interior. Traditionally the tubing is installed primarily in the floor. It may be attached to the underside of the plywood sheathing of a light-frame floor. However, if the tubing is instead embedded within a concrete floor deck, the mass of the concrete stores a great deal of heat. This evens out the spread of the heat, making it more comfortable. It also allows the heating equipment to run at a slow, steady pace, which is energy-efficient and increases equipment life (Figure 13-1).

13-1 *Tubing for radiant heating in position before placing concrete for a floor.* Quad-Lock Building Systems, Ltd.

Another advantage distinct to concrete floors is the option for a *decorative concrete* finish. The top surface of a concrete floor can be covered with any of the conventional finish materials installed over light-frame floors. But in addition, it may be finished with a wide variety of new coloring and texturing methods unique to concrete. These provide an aesthetic option that is difficult to achieve with other types of floor.

Advantages for contractors

The logistics of installation vary so much with the floor system that it is difficult to generalize about the advantages of concrete decks to the installer. Some systems do have logistical advantages for some construction operations.

In most small buildings, concrete decks will generally be more expensive than light-frame decks. The decision to use them hinges primarily on the value the building owner puts on the advantages of concrete.

The deck assembly

In small buildings, concrete decks are usually installed onto concrete walls. Light frame walls rarely have enough strength to hold the weight. Some designers also have concerns about matching up two such different materials in an important structural connection.

A concrete floor is a horizontal plane of concrete resting on the walls. It typically contains large amounts of rebar because it must resist significant, constant weight from above. This requires high bending strength. The rebar installed for this structural purpose is usually located toward the bottom of the concrete. This is where the greatest tensile (or pulling) forces are exerted on the deck.

Concrete roofs may be either flat or pitched. It is typically much easier to create a flat roof with concrete. With cast-in-place systems, creating a pitched roof involves placing concrete on a slope. This can be done, and with some systems it is common and practical. However, it may also be significantly more expensive than a flat roof. Even with precast systems, placing the panels on a slope put high *thrust* (outward force) on the walls. This requires the walls to be designed to endure the extra loads, and that can add cost. However, there are also promising new methods for obtaining a pitched roof without installing concrete on a slope. Chapter 19 discusses this.

It is important to limit cracks closely in concrete decks. In walls, many cracks have little structural effect. Because decks are horizontal, cracks have more potential to grow and either weaken the deck or make it unusable in the cracked location. The usual method of controlling cracks is with a layer of welded wire mesh embedded in the concrete. This is typically installed in the upper part of the floor, well above the structural rebar below.

Different deck systems have very different maximum spans. Almost all of the ones described here span up to 20 feet without difficulty. Some span up to 65 feet. As with any floor system, the longer spans require significantly deeper floors to achieve the necessary strength.

As with frame decks, there are methods of supporting a concrete deck so that it can extend a distance greater than its allowable span. This is done with either interior bearing walls or intermediate cross beams. Each method creates a series of short spans that the deck must bridge instead of one long one. With concrete decks, bearing walls are almost always concrete and built with the same system used to create the exterior walls. Crossbeams tend to be either heavy steel or reinforced concrete. Steel is common for precast decks, because the beam can be set at the same time as the deck panels using the same equipment. Reinforced concrete beams are typical when the floor is cast-in-place. The concrete beam can be created at the same time as the rest of the cast-in-place floor with the same methods and materials (Figure 13-2).

Most of the deck systems involve placing wet concrete on-site. There is sometimes wet concrete used with precast and AAC planks, which arrive at the site

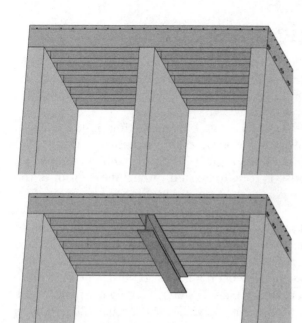

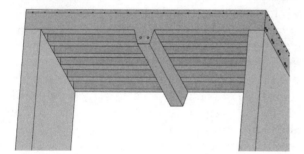

13-2 *Concrete floor supported at an intermediate point with a bearing wall (top), a steel beam (center) and a reinforced concrete beam (bottom).*

cured and solid. In this case, a *bond beam* is often cast around the outside edge of the deck. This is a section of concrete containing rebar on top of the walls. It holds the pieces of the floor in place and ties the walls and floor together structurally (Figure 13-3).

To place the wet concrete, whether it is the entire floor deck or simply a bond beam, it is necessary to install forms around the edge of the wall to hold the concrete in. These are called *rim forms*. The makeup of the rim forms is different depending on whether the walls below are concrete without a facing over them (like block or removable forms walls), or they have another material on the outside surface (like the foam layer of ICF walls or the AAC of AAC walls). If the outside is concrete, the rim form is usually some type of removable form. This may be metal forms, or plywood secured against the outside surface of the walls. The rim form extends up to the height the concrete will be cast to. It will be stripped later to leave an exposed surface of concrete out-

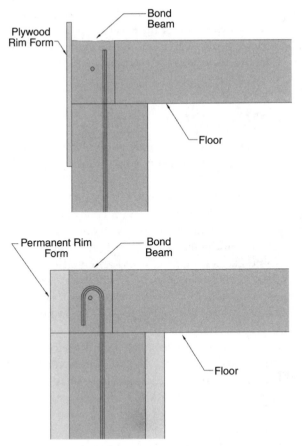

Plywood
Rim Form

Bond
Beam

Floor

13-3 *Rim form and bond beam around a deck above plain concrete walls (top), and above walls with an outside surface of another material (bottom).*

Permanent Rim
Form

Bond
Beam

Floor

side. If the outside is another material, the rim form is made of that material and placed *over* the walls so that the form material's outside surface is flush with the outside of the walls. It will stay in place to provide insulation around the bond beam and keep the outside surface of the building all one material.

Rim forms are held vertical with braces outside or by tying them somehow to the inside.

Code and regulatory

Concrete floors and roofs are covered in the general concrete sections of the International Building Code in the United States and the National Building Code in Canada. Their use in commercial buildings must generally be analyzed by a licensed engineer in accordance with these codes.

They are not, however, covered in the International Residential Code or the residential sections of the National Building Code of Canada. This reflects their lower use in small buildings. Therefore, when they are used in houses, they usually require custom engineering just as in a commercial building. The sellers of some of the systems do provide engineering services or evaluation reports that cover homes. In these cases it may be possible to satisfy the building department that the system is being used correctly without hiring a separate engineer.

Connections

The major attachments to concrete decks are with the exterior walls, with wallboard to create a finished ceiling below, and with the finish flooring or roofing material above.

The connection to the walls is virtually always made with some sort of steel connector or reinforcement that extends between the deck and the walls. The details depend on the exact floor and wall systems involved.

Wallboard may be attached to the underside of most of these deck systems. Again, the details vary. In some cases a plaster product may also be applied directly to the underside of the deck without any intermediate steps.

All common flooring materials are installed over concrete. The tools, fasteners, and adhesives for the job have all been developed and are familiar to most flooring contractors.

Roofing materials for flat roofs are also commonly installed on concrete about as readily as on other materials. On pitched roofs, the common roofing materials are often installed with nails or other fasteners. These may not always connect directly to the surface of a concrete roof. In these cases, additional measures may be necessary.

Architects and engineers

Architects and engineers are generally familiar with concrete decks from use in commercial construction. They may require additional information and preparation time when using some of the newer, less familiar systems.

Energy efficiency

The impact of a concrete roof on a building's energy efficiency is thought to be similar to the impact of concrete walls. Concrete roofs tend to be very airtight, reducing losses from air infiltration sharply. They add yet more thermal mass to the building shell, cutting energy use by evening out the impact of swings in temperature. The roof should always be insulated to insure adequate energy efficiency. If it is constructed of one of the deck systems with a high R-value, this will also contribute to high-energy efficiency.

Concrete floor decks may have an impact as well. Their thermal mass is believed to help even out internal temperature swings in ways similar to the thermal mass in the building shell.

Concrete floors are often instrumental in buildings employing solar heating schemes. The mass of the floor stores heat from sunny times to help warm the interior at other times.

Maintenance and repair

Concrete decks generally require no routine maintenance and seldom if ever need repair. They are particularly resistant to settling or sagging over time. The finishes placed on them wear at the same rates they do when installed over frame decks, and take similar care.

Sustainability

The sustainability attributes of concrete decks are virtually identical to those of concrete walls. The use of concrete decks may contribute to earning Leadership in Energy and Environmental Design (LEED) points for:
- The energy efficiency of the building
- The use of recycled materials, such as fly ash or slag in the concrete or recycled steel in the reinforcement
- The use of local materials, since the concrete almost always comes from within 500 miles and often the steel does, too

Separate from the LEED credits, building with concrete decks is often cited as a method of avoiding the harvesting of trees.

Aesthetics

Because the deck is generally covered, it is the finishes put over it that determine its appearance. As noted previously, most concrete decks are covered with the same finishes used over other decks. This includes gypsum wallboard or plaster underneath. On the top surface of a floor, this includes such conventional materials as wood flooring, carpet, tile, and vinyl. For roofs, it includes rubber membranes, built-up roofing, and roof shingles and tiles. In some cases, special measures may be required to install roofing materials attached with fasteners since these may not easily connect directly to concrete.

In addition, new techniques for texturing and coloring the surface of concrete have led to the development of a whole new set of striking and distinctive finishes for concrete floors. These decorative concrete options are covered in detail in Chapter 22.

Key considerations

Most of the important considerations to keep in mind when building with concrete decks depend on the system used. However, a few apply in many or most cases.

Heating, ventilation, and air conditioning equipment sizing

The possibility that a building's heating, ventilation, and air conditioning (HVAC) equipment will be oversized is even greater when a concrete roof is added. With the entire shell built of concrete, thermal mass will be extremely high and air infiltration will likely be very low. The roof should also be well insulated. If it is, the building should have sharply lower heating and cooling requirements than it would with light-frame walls and roof. The equipment selected by most HVAC contractors will likely be larger than necessary, leading to unnecessarily high equipment cost and inefficient operation. The recommendations in Chapter 5 for dealing with this possibility are even more important when the roof is also concrete.

Air exchange

Likewise, a building with concrete walls and roof will almost always have very low air infiltration. It will therefore be important to have an HVAC design that brings in measured amounts of fresh air. This is also discussed at length in Chapter 5.

A related issue is the potential for humidity buildup. Depending on the climate, warm, moist air may tend to climb inside a building. If the roof is concrete this may lead to excessive moisture under the roof. It may be important for the HVAC designer to include some means of releasing some of this air or recirculating it to the lower sections of the building. This type of issue is routinely taken care of in larger commercial buildings because they tend to have airtight roofs anyway.

Floor rigidity

A concrete floor has less give than a light frame floor. This will make it comparable to the floors in an upscale hotel, for example. Some people will find the surface too hard for some parts of the building. In a home, uncovered concrete floors might prove to be uncomfortable in bedrooms or other rooms where the occupants tend to walk in bare feet.

The solution is typically to install resilient flooring over concrete in key rooms. For example, most hotels use padded carpet in the guest rooms.

14

Composite steel joist floors and roofs

Composite steel joist decks are a cast-in-place system with some novel features. Construction begins with installation of special joists on top of the walls. The joists are made with welded steel bars in a truss pattern. A concrete slab is cast on top with only standard plywood as the form below and only a layer of welded wire mesh as the reinforcement (Figure 14-1).

The elimination of some materials and steps make the system relatively economical even on small buildings. The completed floor includes a wide-open plenum for quick and flexible installation of ductwork, plumbing, and wiring. Clear spans of over 40 feet are possible when a sufficiently deep joist is used. The system includes no insulation, but that can be added. Although flat roofs are possible, composite steel joists are not readily adapted for use in pitched roofs.

Composite steel joist floors have gradually grown in mostly larger buildings since the 1970s. When the buyers of small concrete buildings became interested in having concrete floors as well, sales of the system picked up in that market, too. Today they are one of the single most used systems for creation of concrete floors on small buildings in North America.

Using composite steel joists requires little initial investment because there is no special equipment to buy. Installation of the hardware is readily learned. The concrete work is the same as conventional elevated flatwork, for which crews are widely available.

14-1 *Worker setting roll bars in position for composite steel bar joist floor.* Hambro
Structural Systems

History

The composite steel joist deck system first appeared in the 1970s in Canada, where it was invented and designed. It was used primarily in mid-size commercial buildings. Use of the system grew steadily over the decades.

In the mid-1990s, buyers of small concrete buildings began to request concrete floor decks to go with their concrete walls. This was particularly common in high-end homes. Contractors tried a variety of concrete floor systems for the job, most of them adapted from commercial and larger-scale construction. The composite steel joist system, designed with a somewhat smaller building in mind, proved more suited to these very small-scale applications.

Use has spread steadily throughout North America. The advantages of this system have made it one of the most used for concrete floors in small concrete buildings.

Currently, five North American plants supply the product. One is in Canada, two are in the United States, and two are in Mexico. Occasionally foreign plants also ship the same or a similar product for use in North America.

Market

The composite steel joist system has long been used in commercial buildings. As some of the smaller commercial buildings have come to be constructed with concrete walls, it has often been chosen to create the floors and flat roofs in these. Common uses have been in offices, hotels, and apartment and condominium buildings.

More recently, buyers of concrete homes have come to request concrete decks in place of light frame to create a "total concrete shell". Probably 5 percent to 10 percent of all concrete homes with floor decks now use concrete, and that appears to be growing. To fulfill this request, many contractors have chosen to use the composite steel joist system.

Advantages to the owner

The general advantages of concrete decks apply to composite steel joist decks, including:

General advantages:
- Resistance of the structure to fire
- Resistance to rot, mold, mildew, insects, and vermin
- Strength and resistance to vibration and settling
- Added structural integrity to the building
- Durability

Floor advantages:
- Reduced sound transmission between floors
- Extra thermal mass inside
- Rigidity and lack of vibration and "bounce"
- Possible embedment of radiant heating tubing
- Possible creation of a decorative concrete top surface

Roof advantages
- Thermal mass and air tightness for energy efficiency and comfort
- Reduced sound transmission from outdoors
- Resistance to wind damage

The system does not include insulation. That would need to be added to make an energy-efficient roof.

Advantages for contractors

The composite steel joist floor is one of the more easily adapted to small building construction. There is no special equipment and no structural rebar to install. There is no shoring. All pieces can generally be lifted into place manually.

When the wall is complete, there is high flexibility for running ductwork, plumbing, and wiring within it.

The joists are ordered to length and are not usually easy to modify in the field. They are specially ordered from the manufacturer. It is also necessary to obtain crossbars called *roll bars* from the manufacturer and return these when construction is done.

Components

The key component of the system is the joist. It is a truss-style joist made of welded steel bars, with steel flanges along the top and bottom.

The bottom flanges are L-shaped steel angle. They are positioned back-to-back to create an inverted-T. They are welded to the bars. Together they create the *bottom chord* of the truss.

The top flange is the *top chord* of the truss. It is welded to the bars also. It is deformed into an "S" shape along the top to embed solidly in the cast concrete. It has periodic holes along it for engaging the roll bars.

A fixture called a *shoe* is welded to each end of the joist at the top chord. The shoes are attached at the factory after the joists are cut to length.

Along the bottom chord of a joist are welded small fittings called *clips*. These are arranged so that the ends of roll bars can fit into them.

The joists are available in depths of 8 to 24 inches, in increments of 2 inches. The length is custom cut for the project, but it generally ranges from 20 feet to 43 feet. They are produced with a slight camber. The camber ranges from 0 inches to ¾ inches for joists under 20 feet long, up to 1½ inches to 2 inches for joists over 30 feet long.

The *roll bars* are steel crosspieces with notched ends. The notches fit securely into the holes along the top chord of the joists. Their standard length is a little over 4 feet. Telescoping roll bars are also available to fit closer spacings of the joists.

Standard 4 feet by 8 feet sheets of plywood are used as formwork. They form the underside of the concrete slab.

Standard welded wire mesh is used for reinforcement and crack control.

Floor assembly

The shoes on the end of each joist rest on the exterior walls. The joists are normally spaced slightly over 4 feet on center. They may be spaced closer as the floor layout requires (Figure 14-2).

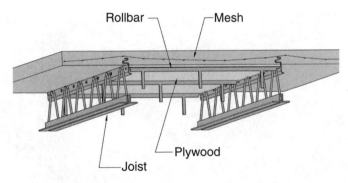

14-2 *Cutaway view of a composite steel joist floor.*

The concrete slab is 2½ inches to 3¾ inches thick. The thickness required depends on the span of the floor. Greater distances require greater strength, and therefore more concrete. The slab extends over the exterior walls, so that its outside edge is flush with the outside surface of the walls.

Embedded in the slab is welded wire mesh. It extends over the entire area, from wall to wall. It is draped over the top chords of the joists.

Ducts, piping, and plumbing can run between and through the joists in the plenum space the joists create.

Insulation is optional. It may be installed between the joists much as it is installed between wooden joists or rafters. Board insulation may also be installed in sheets to the underside of the joists.

If the underside is finished, this is almost always with conventional gypsum wallboard. Three-quarter inch steel hat channel is attached to the bottom side of the joists, running perpendicular to them. It is held tight with wire to the bottom chord. The wallboard is screwed to the hat channel (Figure 14-3).

Cost

Subcontractors typically charge $5.00 to $9.00 per square foot for installed floors using the system. The square footage includes any openings in the floor. The price includes all materials, all labor, equipment, and the subcontractor's markup.

Estimated costs, without markup, for a typical floor project are listed in Table 14-1.

Costs vary with the size of the project and the geographic location in which the work is performed. The cost per square foot tends to be lower on larger projects because the setup labor is spread over a larger floor area. Location is important because labor rates vary from place to place. In addition, projects far from a manufacturing plant result in higher charges for the joists to cover the longer shipping.

14-3 *Hat channel attached to the underside of joists.* Hambro Structural Systems

Also important is the span of the floor. Longer spans require deeper joists and thicker slabs. These increase materials cost.

Code and regulatory

The major manufacturers of this system have evaluation reports from the U.S. and Canadian building code bodies. They also have staff engineers who will perform the engineering of the floor for each project.

Table 14-1. Approximate Costs of Composite Steel Joist Floor Construction.

Item		Cost per Square Foot*
Labor		$2.50
Set joists and formwork	$1.75	
Place concrete	$0.75	
Materials		$3.25
Joists (includes roll bar rental)	$2.25	
Concrete	$0.75	
Mesh	$0.25	
Pump rental		$0.35
Total		$6.10

*All costs in U.S. dollars per gross square foot floor area.

Most building departments will require engineering on all buildings. However in some areas where the product is frequently used, they may be satisfied on small projects without custom engineering.

Installation

It is important to plan the layout of the floor in advance of ordering the joists. The positions of the joists must not conflict with features such as openings.

The joists are ordered to exact length. The factory produces them to length and attaches a shoe to each end.

The workers at the site rest the ends of the joists on the bearing walls. On a concrete wall, each shoe must have a bearing of at least $3\frac{1}{2}$ inches. They check for level and place shims under the shoes as necessary to achieve it.

The joists are set slightly over 4 feet on center from one another. They may be closer when the floor layout requires it.

Roll bars go between adjacent joists. Where joists are spaced closer than 4 feet, telescoping roll bars are fit between them. One roll bar is generally installed every 21 inches on center along the length of a joist. There are also roll bars near the front and back wall. An occasional roll bar also goes into the clips along the bottom chords of the joists, to connect the bottoms of the joists to one another. This prevents them from "rolling" to one side or the other throughout the installation.

Sheets of plywood go between the joists. They rest on top of the roll bars. Either $\frac{3}{8}$-inch or $\frac{1}{2}$-inch plywood is used. It is oiled to make stripping easier.

The crew drapes welded wire mesh over the top chords of the joists. The sheets have minimum overlap requirements for one another, and overlap the bearing walls as well. A rim form goes around the outside of the building.

Concrete specifications may vary, but typically it must be not less than 3000 psi, with aggregate size no larger than $\frac{3}{4}$-inch and slump between 4 inches and 5 inches. It is placed so that the top surface is at least 1 inch over the top edge of the top chords. It must be consolidated thoroughly.

Typically after one day the crew can remove the roll bars and plywood below. It is common to work on the floor within about two days (Figure 14-4).

Connections

The concrete walls typically have vertical rebar extending upward from their top edges. These will extend through the slab and up further. They will later connect to the walls that will be formed up above (Figure 14-5).

Stairs are usually installed by means of a buck. Before floor construction, a buck is constructed to create a rough opening for the stairway. It is braced in position between joists. Typically anchor bolts are bolted to the buck and their bent ends extend into the slab space. After concrete is cast, the bolts em-

14-4 *Worker removing roll bars and plywood.* Hambro Structural Systems

bed in the slab and the bracing is removed. The staircase may be fastened directly to the buck.

Floor coverings attach as they do onto any concrete floor.

As noted previously, to create a ceiling workers attach steel hat channel to the joists, and attach wallboard to the hat channel.

Architects

The manufacturers have architects they can recommend who are familiar with the system. However, learning the essentials is straightforward and new archi-

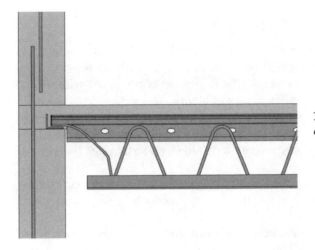

14-5 *Connection of floor to concrete wall.*

tects adapt easily. The manufacturers can provide design materials and help them get up to speed.

Engineers

The manufacturers have in-house engineers who do the structural design of the floors. They provide stamped plans to the building department. If needed, they will coordinate with the engineers designing the rest of the building.

Training

In most projects installation is given to crews already familiar with the system. The manufacturer or its local distributor can provide references to experienced crews.

Because the floor is a simple assembly, new crews often find they can learn installation simply from the manufacturer's manual. The manufacturer also has field representatives who are available to visit the job site of a first-time crew to assist and instruct. This is highly recommended.

Maintenance and repair

There is normally no maintenance or repair required. Like concrete floors in general, composite steel joist floors are resistant to shifting or sagging. They are not expected to need adjustment or repair for the life of the building.

If the joists are left exposed they are frequently painted. In this case, they may need periodic repainting.

Energy

Used as a roof, the system should provide good air tightness and significant thermal mass. The thermal mass will be a little less than with some other concrete systems because composite steel joist floors are constructed with a relatively slim slab and without regular concrete beams.

Insulation will generally be necessary to make the deck a suitable energy-efficient roof. This may be installed between the joists just as it is installed between wood joists. It may also be installed as sheet material over the bottom of the joists.

As a floor, the system may be called on to separate heating or cooling zones. In this case, it could be insulated as it would if it were used as a roof.

In any case, the floor should help even out and reduce the heating and cooling requirements of the building because of the additional thermal mass it provides inside the building. If a solar heating scheme were used, the floor could provide significant heat storage.

Sustainability

The usual sustainability properties of concrete apply to this system. It may contribute to earning Leadership in Energy and Environmental Design (LEED) points for energy efficiency, use of recycled materials, and use of local materials, as discussed in Chapter 13.

Separate from the LEED credits, building with concrete decks is often cited as a method of avoiding the harvesting of trees.

Aesthetics

As noted, any conventional floor covering can be attached to a concrete floor. In addition, the full range of decorative concrete finishes is available to create striking, distinctive floor surfaces that would be more difficult on other types of floors. The details of decorative concrete floors are in Chapter 22.

If the underside of the floor is finished to create a ceiling, this is normally done with conventional wallboard, as previously noted.

Key considerations

During installation it is important to follow procedures that do not overstress the joists. They will carry all the weight of the concrete, people, and equipment until the concrete cures. Heavy equipment, used for such tasks as finishing wet concrete, can cause problems. In addition, the roll bars alone are not designed to have the load carrying capacity of the joists.

All roll bars and joists need to be secure so that they remain in position through the pour. If the slab thickness will exceed 3½ inches, the roll bars must be spaced closer together than usual. The plywood sheet should be tightly butted along their edges. When using ⅜-inch plywood, it may be acceptable to overlap the sheets.

If the system is used as a roof and the walls are also concrete, the building is likely to be extremely airtight. The HVAC contractor or engineer should be consulted about measures to maintain proper air intake and humidity control.

When used as a roof, it will be very important to insulate. However, once the roof and walls are insulated, the heating and cooling load of the building will likely be sharply lower than in most light frame buildings. Take steps to insure that the HVAC equipment is properly sized for such an energy-efficient building. Previous chapters discuss these steps.

Availability

Manufacturers will ship the joists and roll bars almost anywhere. However, shipping costs are higher farther from the factories. Currently there is one factory in Quebec Province, Canada, one in Washington State, one in Florida, and two in Mexico.

Crew availability

Crews experienced with the product are throughout North America, with more nearer the locations of the plants. However, installation of the system is easily learned, so having a crew that is already experienced is not usually critical. The manufacturer can help locate an experienced crew or assist a new one.

Support

The manufacturers are set up to provide a wide range of support for the system. They have local representatives in some areas. There is support staff at the plants themselves.

Current projects

The local plant or distributor selling the product should be most helpful in locating projects currently underway that might be visited. These can be located through the manufacturers.

15

Insulating concrete form floors and roofs

Like insulating concrete wall forms, insulating concrete forms (ICFs) for decks are made primarily of plastic foam and remain in place after the concrete cures to act as insulation. They create a "half-sandwich" with reinforced concrete on top and foam underneath (Figure 15-1).

The insulating concrete forms provide flexibility because the material can be easily cut and shaped. They result in a highly insulated deck, which is especially useful for roofs. Fastening surfaces also make attachment of finishes on the underside relatively easy.

In a few short years, ICF decks have become a favorite in small concrete buildings. Their flexibility makes them adaptable to buildings of any size and shape, and to roofs that are either flat or pitched. The forms are believed to be sold in the amount of several million square feet of forms per year and to be growing 10 percent to 20 percent per year.

Because the forms stay in place, the cost of an ICF deck includes the cost of a set of forms. It also requires the cost of shoring to hold the forms level during construction.

History

Insulating concrete deck forms were invented in Italy in the early 1980s. At that time they included no steel members to stiffen them. They were made of pure foam.

The European producers added the steel members in 1999. In the same year they began production and sales in the United States. Local companies

Color Plate I

Home with concrete brick veneer

Home with concrete block veneer

Fiber cement lap siding

James Hardie ® Siding Products

Fiber Cement Plank siding

Stucco finished home

Color Plate III

Removable forms house with a form-liner finish

Concrete roof tiles

Thin block veneer on tilt-up wall panel

Manufactured Stone siding

Color Plate IV

Precast Walls with thin brick veneer

Acid stained concrete countertop

Color Plate V

Stone Soup Concrete

Concrete Countertop and farm sink

Floor Seasons. Inc.

Acid Stained and textured floor

Color Plate VI

Acid stained and cut floor

Manufactured stone wall

Manufactured stone fireplace

Pavers by Ideal

Segmental retaining walls

King's Materials

Concrete pavers

Color Plate VIII

Stamped and colored exterior flatwork

15-1 *Workers installing ICF deck forms.* INSUL-DECK LLC

followed with their own versions of the product soon after. There are now several plants producing the product around the United States.

Initial sales were for very small buildings, especially single-family homes. Because structurally the decks are reinforced concrete, they are fully capable of creating large and high-strength floors and roofs as well. As the method became better established it sold to a growing number of larger projects as well.

Market

Insulating concrete form decks are often installed on buildings with ICF walls. This is logical. An owner that demands the energy efficiency of foam-insulated walls is likely to look for the same in the roof. In addition, the wall crew is set up to work with foam and can readily build the decks as well. Note, however, that ICF decks can be installed on any type of reinforced concrete wall.

Use of ICF deck forms is spread across the United States and Canada. The applications are also diverse, including every building type ICF walls are used for. However, at this time most of the product use is in commercial construction and multi-family residential buildings. It is also used for high-end single-family homes. Insulating concrete form decks are an insulated product, so they are used primarily on buildings that require controlled environments, such as apartments and condominiums, theaters, hotels, offices, storm centers, cold storage, certain agricultural and industrial buildings, and others that house people or sensitive processes.

Advantages to the owner

The general advantages of concrete decks apply to ICF decks, including:
  General advantages:
  • Resistance of the structure to fire
  • Resistance to rot, mold, mildew, insects, and vermin
  • Strength and resistance to vibration and settling
  • Added structural integrity to the building
  • Durability
  Floor advantages:
  • Reduced sound transmission between floors
  • Extra thermal mass inside
  • Rigidity and lack of vibration and "bounce"
  • Possible embedment of radiant heating tubing
  • Possible creation of a decorative concrete top surface
  Roof advantages:
  • High thermal mass and air tightness for energy efficiency and comfort
  • Reduced sound transmission from outdoors
  • Resistance to wind damage

Insulating concrete form decks also have a high R-value. The insulation and the cost of the insulation are included in an ICF deck. The ease of cutting and modifying the forms makes it practical to use the system in irregular layouts, including odd shapes and pitched roofs.

Advantages for contractors

Insulating concrete deck forms are relatively easy to work with in the field. They are light and easy to lift into position. They are readily cut to fit as needed. It is also relatively easy to cut deep notches in them to create forms for extra beams that may be needed on some floor designs.

Shoring is required beneath the forms to support them before the concrete hardens. The workers must take care not to step into the depressions in the form (called *beam pockets*) where concrete beams will be formed. The thickness of the forms there is less, and is not sufficient to support a person's weight.

Components

Insulating concrete deck forms come in *sections* that are two feet wide. They are supplied cut to length. The sections are designed to be placed side-by-side to create the forms for the deck (Figure 15-2).

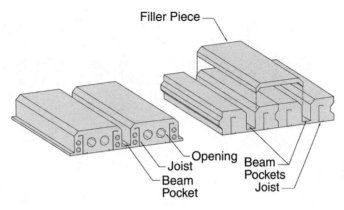

Filler Piece

Opening
Joist
Beam
Pocket

Beam
Pockets
Joist

15-2 *Two styles of ICF deck forms.*

The sections have a profile with thinner parts that run the length of the section. These create beam pockets for the deck. Some brands of form also have cylindrical cavities that run the length of the section.

Embedded in the foam of each section are two light-gauge steel members that run the length of the section. The members give the section stiffness and provide a point for attachment of wallboard for the ceiling below.

The foam of the sections is expanded polystyrene.

The sections are available in different thicknesses. The thicker sections have deeper beam pockets. Beam pockets of different depth are required for different strengths of deck. Some producers also provide special filler pieces that fit over the thicker portion of the section to build it up. Adding these makes it possible to create a deeper beam pocket, as needed in the field.

The other materials needed to create the deck are conventional concrete, steel reinforcing bar, rebar chairs, and welded wire mesh.

The deck assembly

The sections extend the depth of the building, forming a foam deck with a beam pocket every two feet on center (Figure 15-3).

Concrete fills the beam pockets and extends above them to create a concrete slab above that is typically about 3 inches thick.

Rebar runs inside the beam pockets, extending from one wall to the other. These bars lap other bars that extend out of the walls and into the beam pockets. Steel mesh extends throughout the concrete slab. It is generally centered in the depth of the slab.

Chases may be cut in the foam as needed for ductwork, pipes, and wiring. These may also run in the precut cylinders in the foam.

The light-gauge steel members run along the bottom of the sections.

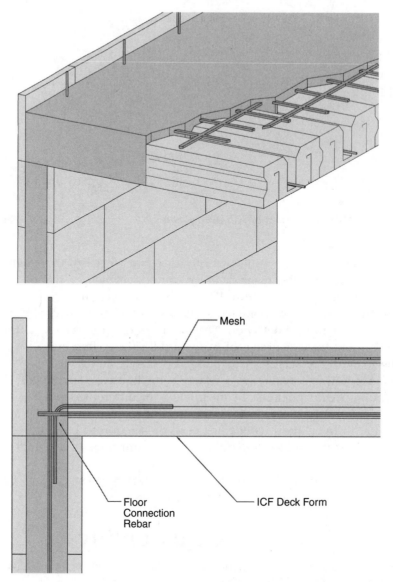

15-3 *Cutaway view of an ICF deck (top) and connection to walls (below).*

Cost

The installed cost of an ICF deck varies widely depending on the particular project. Recent quotes from subcontractors range from about $8.00 to $18.00 per square foot. This is total installed cost, including the subcontractor's markup.

The installed cost is often more than that of some other concrete floor systems. However, those do not include insulation and may not be as readily

Table 15-1. Approximate Costs of ICF Floor Construction.

Item		Cost per Square Foot*
Labor		$3.25
Set shoring and formwork	$2.50	
Place concrete	$0.75	
Materials		$5.35
Forms	$3.75	
Concrete	$1.00	
Rebar	$0.35	
Mesh	$0.25	
Shoring rental		$0.50
Pump rental		$0.35
Total		$9.45

*All costs in U.S. dollars per gross square foot floor area.

adapted to odd layouts, changes in the field, and pitched roofs. In small buildings it is usually more than the cost of a light frame deck, but that may not have the degree of insulation or the advantages of concrete.

Estimated costs for a typical floor project, without markup, are listed in Table 15-1.

Costs per square foot are generally lower for large projects and higher for small ones. Short lengths involve more labor per square foot. In addition, setup charges are spread over a smaller floor area.

Pitched roofs will be closer to the high end of the cost range. This is because of the extra work involved in setting pitched shoring, creating ridge beams, and placing concrete on a slope.

Code and regulatory

Almost all ICF floor installations are custom engineered for the individual project. The buyer retains an engineer to do this work. Some of the manufacturers have extensive manuals and on-staff engineers to assist.

Insulating concrete form floors are considered to be a type of cast-in-place concrete floor. They are therefore covered by the general concrete provisions of the International Building Code, in the United States, and National Building Code, in Canada. They are not included in the International Residential Code or the residential section of the National Building Code.

Like all foam used in construction, the foams of ICF deck systems are required to meet certain minimum fire resistance standards. The manufacturers can provide documentation verifying that their products comply.

Installation

Insulating concrete form decks are usually installed on ICF walls. They may, however, be installed on almost any concrete walls.

Floors

The layout of the floor is planned out in advance of building the walls below. It is necessary to install extra rebar in the walls, and these must line up with the locations of the beam pockets of the floor. The beam pockets are positioned so that they do not interfere with openings needed in the floor.

The floor forms are generally installed after the concrete of the walls below has cured. The shoring is erected first to support the forms as they are set in place. The shoring runs in lines from sidewall to sidewall of the building. One line of shoring must be in place every six feet or less, with an additional line of shoring up against the front wall and one against the back wall

The form sections are usually set on top of the walls below. The ends of the sections bear about two inches on the walls. On an ICF wall, this puts the ends of the sections over the inner layer of foam on the walls, so that foam bears only on foam. It may be unacceptable for the foam to bear on the structural concrete of a wall because the foam would encroach on the concrete wall space and create a line of weakness in the walls. To avoid this on block or plain concrete walls, sections can be set inside the walls and butted up to them and rest solely on the shoring.

There is rebar pre-installed in the walls so that one bar lines up with each beam pocket in the floor forms. These *floor connection rebar* are in addition to the rebar required for wall reinforcement, which will continue upward into the next story of walls above. The floor connection rebar are bent so that they extend into the beam pockets, usually by about two feet.

Additional structural rebar are set in the beam pockets on chairs. They extend the entire length of the pocket and lap the floor connection rebar at each end. Welded wire mesh is set above the foam on chairs and rim forms are installed around the edge of the building (Figure 15-4).

With all forms and steel in position, the concrete crew places concrete like any elevated flatwork. The placement is organized to avoid placing undue stress on any one point of the forms. Concrete is not allowed to pile up in one spot. The concrete is finished like any flatwork and allowed to cure.

Roofs

Flat roofs are installed almost exactly like floor decks. The only significant difference is that the structural rebar in the walls will not be extending up past the floor to connect to another set of walls above.

Pitched roofs require pitched shoring and a ridge beam. Pitching the shoring requires that each line of shoring be at a different height to hold the forms at

15-4 *Forms and reinforcement in position and ready for concrete.* Technologies

the correct roof pitch. The forms must also be secured in position to prevent them from sliding.

The ridge beam is formed much like a center beam for a floor, with a custom-made plywood form. The ICF sections are cut short so that they reach up to the edge of the beam form. Rebar goes in the beam form.

When concrete is cast it fills the center beam forms as well as the ICF forms, creating a single integrated structure (Figure 15-5).

The concrete must be placed on a slope. This requires extra skill and attention from the concrete crew. Typically they specify a low-slump concrete. They continuously "push" the concrete up to counter its tendency to slide down. Most crews find it difficult to place concrete on a pitch greater than 4-in-12. However, roofs as steep as 6-in-12 have been successfully constructed.

Connections

The decks are connected to concrete walls with extra rebar termed "floor connection rebar," as already discussed.

Stairs are usually installed by means of a buck. The foam must be adjusted to position a beam close along either side of the buck, and crosswise along its front and back. These beams are outfitted with rebar. The anchor bolts of the buck extend into these beams.

Floor coverings attach as they do onto any concrete floor.

Wallboard fastens with conventional screws to the steel members along the bottom of the forms.

15-5 *Placing concrete on a sloped ICF roof.* INSUL-DECK LLC.

Architects

Insulating concrete form floors are typically custom designed for each project, with assistance from the manufacturer. Architects have generally proven to be receptive to the product.

The design must take into account where openings will occur so that beams can be placed around them. However, this is not much different from other floor systems, which generally also have joists or beams. Insulating concrete form floor decks have good flexibility to be cut and shaped as needed. Spans of up to 36 feet have been accomplished with conventional reinforcement. The longer spans require deeper floors.

Engineers

These decks are almost always custom engineered for each building. Typically the buyer has hired an engineer for the rest of the building and that same engineer designs the floor.

Engineers typically adapt readily to the system. Structurally ICF floors are simply reinforced concrete, which is widely familiar. Some of the manufacturers provide extensive engineering documents and have engineers on staff to consult with the building's design engineers.

Training

Assistance to first-time crews is often available from the local sellers of the ICF deck forms. The forms are sold by many building product distributors. Most of these are also sellers of ICF wall forms. They can be found by the methods described in Chapter 7.

Typically the forms manufacturer consults with first-time crews. The manufacturers also provide manuals and informational CDs. However, no formal training classes currently exist.

Maintenance and repair

There is normally no maintenance or repair required. Like concrete floors in general, ICF floors are resistant to shifting or sagging. They are not expected to need adjustment or repair for the life of the building.

The foam should be covered below with ordinary gypsum wallboard. If it is not, it may be susceptible to damage over time from ultraviolet rays or impact.

Any coverings or finishes placed on top of the floor or roof may require maintenance or periodic replacement. Consult the floor- or roof-covering manufacturer for information.

Energy

Used as a roof, the system provides high air tightness, thermal mass, and R-value. The R-value will depend on the thickness of the foam used. Consult the manufacturer for the estimated R-value of a particular floor.

Used as a floor, the system may be called on to separate heating and cooling zones. It should be highly effective.

In addition, an ICF floor should help even out and reduce the heating and cooling requirements of the building because of the additional thermal mass it provides inside the building. If a solar heating scheme were used, the floor could provide significant heat storage as well.

Sustainability

The usual sustainability properties of concrete apply to this system. It may contribute to earning LEED points for energy efficiency, uses of recycled material in the concrete and rebar, and use of local materials. The systems do not typically use significant recycled content for the plastic foam in the forms.

Separate from the LEED credits, building with concrete decks is often cited as a method of avoiding the harvesting of trees.

Aesthetics

Any conventional floor covering can be attached to a concrete floor. In addition, the full range of decorative concrete finishes is available to create striking, distinctive floor surfaces. The details of decorative concrete floors are in Chapter 22.

If the underside of the floor is finished to create a ceiling, this is normally done with conventional wallboard.

Key considerations

Generally speaking, this should be treated as a "commercial" product, even when used in a home. It must be designed by an engineer. Shoring should be designed by an engineer. Crews and construction methods used should be those from commercial construction.

Adequate shoring is critical. A common mistake is not adequately side-bracing the shoring, so that it tips. Site-built wooden shoring is also risky. The wood may have defects, and the shoring is often of uncertain strength.

Other potential mistakes include using tall, narrow shoring beams that may twist to one side. Also, shoring beams butted end-to-end need to be supported beneath the joint.

It is important to use a crew experienced with elevated flatwork. Errors can have big consequences, so it is wise to hire a crew that is least likely to make mistakes.

Installing an ICF roof over concrete walls will likely produce an extremely energy-efficient building. The precautions listed in Chapter 13 about sizing heating, ventilation, and air conditioning (HVAC) should be followed to avoid installation of equipment that is larger than necessary, costly, and inefficient. Likewise, the building will probably be very airtight, so the precautions recommended in Chapter 13 regarding provision of adequate ventilation should be followed.

Availability

ICF deck forms are sold throughout North America. Many local distributors of ICF wall forms also sell the deck forms. Locate these distributors by the methods listed in Chapter 7.

Crew availability

Installing the forms and placing the concrete are usually done by two separate crews.

On a building with ICF walls, the form installation is almost always performed by the ICF crew. If the walls are of some other type, a forms crew or carpentry crew can install the forms. If the crew is not experienced with the system they should seek support from the manufacturer or the local distributor who provided the forms. To find a suitable crew, ask for references from the local distributor.

An elevated flatwork crew places the concrete. These crews are widely available across North America. The local forms distributor may know crews that could do the work.

Support

The local distributor that sold the forms can provide considerable assistance, and can direct the user to other sources of help. The distributor is often available to come to the job site. The manufacturer also has staff that can provide engineering assistance and help in many other matters. On some general issues, the Insulating Concrete Form Association (www.forms.org) may be of help.

Current projects

The local distributor selling the product should be most helpful in locating projects currently underway that might be visited.

16

Precast plank floors and roofs

Precast planks create rigid concrete decks quickly. They arrive at the job site cut to length and already reinforced. Cranes set them in position on the walls, completing a floor or roof structure in a few hours (Figure 16-1).

Few wet materials are used on-site. Installation time is very short. Spans of up to 60 feet are possible with sufficiently thick planks. The planks are available with or without insulation. They may also have hollow cores, where it may be possible to run some ductwork, plumbing, and electrical wiring.

The planks are ordered in advance. They are ordered to length and are not ordinarily cut or adjusted in the field. Precast plank decks are especially efficient when used in conjunction with precast walls. Typically a crane is used to install all components, and this may often be done all with one trip from the crane.

Pitched roofs are not normally constructed with precast plank because of the difficulty of holding the heavy parts in place on a slope. Precast plank has become a popular method of building a concrete deck on concrete walls quickly and with minimum site work. Tens of millions of square feet of precast plank are sold for use in floors each year. The amount used in small buildings is unknown, but is a smaller part of the producer's market.

History

Precast floors were used in some of the very first precast buildings in the late 1800s in the United States. Hollowcore concrete planks appeared in the United States and Canada in the 1950s. Their profile with empty cylindrical cores was an efficient way to create a slab of concrete that is high in strength.

After energy prices jumped in the 1970s, versions of planks appeared with a layer of foam insulation sandwiched between two layers of concrete.

16-1 *Setting precast floor plank into position.* The Spancrete Group, Inc.

Precast planks originally sold well for use in large buildings. This remains their primary market today. However, in the late 1990s, buyers and builders looking for practical methods of creating concrete decks on small concrete buildings discovered them. They found that the product is readily adaptable to a small building.

Market

About 30 percent of the sales are believed to be for use in residential buildings, including apartment or condominium buildings and some homes. The rest are in commercial, industrial, and institutional construction. The speed of construction has particularly made the product attractive to the retail and hotel and motel markets. Speed is valuable there because stores or hotels that can open early may bring in hundreds of thousands of dollars of additional revenue.

Sales are spread broadly throughout the United States and Canada. The product has particular appeal in certain types of areas. In cold regions construction can proceed during low temperatures, because of the limited use of wet materials. In high-wind areas, the strength of the deck is attractive.

Advantages to the owner

The general advantages of concrete decks apply to precast decks, including:

General advantages:

- Resistance of the structure to fire
- Resistance to rot, mold, mildew, insects, and vermin
- Strength and resistance to vibration and settling
- Added structural integrity to the building
- Durability

Floor advantages:

- Reduced sound transmission between floors
- Extra thermal mass inside
- Rigidity and lack of vibration and "bounce"
- Possible embedment of radiant heating tubing
- Possible creation of a decorative concrete top surface

Roof advantages:

- High thermal mass and air tightness for energy efficiency and comfort
- Reduced sound transmission from outdoors
- Resistance to wind damage

Added to this are some particular advantages of precast plank. As noted above, the speed of construction is high. The planks can be insulated, adding to the energy efficiency of a roof, or permitting zoning of heating and cooling between floors. The clear spans possible with the plank are higher than almost any other floor system suitable for small buildings.

Advantages for contractors

Installation of precast planks can be fast and predictable. There are few steps in the field and few special tools. As a result, there is also little new that a contractor must learn to use the product.

The logistics are different from those of many products for small buildings. The planks may need to be ordered well in advance and are not ordinarily modified or adjusted in the field. A crane does much of the work of installation.

Components

The planks are available in different dimensions. They may also be ordered with or without a layer of insulation.

Most planks have a tongue-and-groove or shiplap feature along the side edges so that they can overlap when the planks are set alongside one another.

Uninsulated planks are virtually always produced with hollow cylinders running the length of the plank. These are called *cores*. Plank with cores is sometimes referred to as *hollow-core*. Standard thicknesses include 4 in., 6 in.,

8 in., 10 in., 12 in., and 15 in. The thickest planks can span up to 60 feet in some situations. Four-inch thick plank is usually supplied in widths of 48 inches or less. Thicker planks are generally 48 inches to 96 inches wide. All are supplied cut to length.

Hollowcore planks are reinforced with prestressed steel cables. The cables extend the length of the plank.

Insulated planks are sometimes called *sandwich panels* because they consist of a layer of foam between two layers of concrete. The simplest of these have two solid concrete layers approximately 2 inches thick, with no cores in the concrete. Connecting the concrete layers are steel trusses that run the length of the plank. The top chord of the truss is embedded in one of the concrete layers, and the bottom chord is embedded in the other to hold them a constant distance apart and act as reinforcement. The foam fills the space between the concrete layers (Figure 16-2).

Some sandwich panels have a two-inch layer of concrete on one face, and a full hollow-core plank on the other, with the layer of foam in between. Steel trusses connect the concrete layers. This variety of plank provides insulation combined with high strength for spanning long distances.

The deck assembly

The planks are butted side-by-side on top of the walls below. Their ends rest on the walls. The planks may extend to the exterior surface of the concrete walls. As an alternative, they may only bear on the walls only 3 to 4 inches.

The planks are fixed into position by various methods. They may be bolted or welded to the walls and to each other. Alternatively, a bond beam cast around the perimeter of the walls may hold them.

On top of the planks is sometimes an extra layer of concrete to level the top surface. This is usually 1 inch to 2 inches thick. Ducts, piping, and plumbing may run in the cores.

If the underside is finished, this is usually with a plaster product applied directly to the planks. The top may be finished with virtually any conventional floor covering. If the top has an added layer of concrete, this may also be finished with the decorative concrete techniques discussed in Chapter 22.

Cost

The total installed cost of uninsulated, precast hollow-core plank is approximately $7.00 to $9.00 per square foot. This is the price to the general contractor, including the subcontractor's profit. However, price can vary significantly. It is generally similar to the cost of other comparable concrete floor systems. In small projects it may be higher than wood frame, but it will also have the benefits of concrete.

16-2 *Sandwich plank before insulation is installed (left) and after (right).* Dukane
Precast, Inc.

16-2 *(continued)*

Table 16-1. Approximate Costs of Uninsulated Precast Floor Construction.

Item		Cost per Square Foot*
Labor		$1.50
Materials		$4.75
Planks	$4.50	
Misc.	$0.25	
Crane rental		$0.35
Total		$6.60

*All costs in U.S. dollars per gross square foot floor area.

Estimated costs for a typical uninsulated floor on a smaller project with panels welded in position are listed in Table 16-1.

Costs tend to be lower for large projects and higher for small ones. Insulation may add $3.00 to $5.00 per square foot. Most of that is to pay for the more expensive insulated panel. A concrete topping on the panels may add about $1.00 per square foot. Unique or specially shaped planks will be more expensive and add to cost. If the planks come from a plant that is far from the job site, shipping may increase their cost. Also important is the span of the floor. Longer spans require thicker planks that carry a higher price per square foot.

Code and regulatory

Precast decks are covered by the general sections on concrete construction in the International Building Code, in the United States, and National Building Code (NBC), in Canada. They are not specifically covered in the International Residential Code or the residential sections of the NBC. Virtually all installations require custom engineering by a licensed engineer.

The major plank manufacturers have evaluation reports from the United States and Canadian building code bodies. They also have staff engineers who can perform the engineering of the floor for small projects and assist the engineer on large ones. This supporting information smooths the way for acceptance of the system by local building departments.

Installation

Crew size depends heavily on the size and scope of the job. Crews on very small jobs like houses are typically three or four workers. On a larger project there may easily be six or seven workers. Usually, one supervisor is highly skilled and trained in precast installation. The others also require training to insure safety and prevent damage on the job.

The manufacturer provides shop drawings that show the location of each plank on the walls. Before installation, the tops of the walls may be marked at the plank locations to insure precise placement.

Level deck

Typically installation begins at one corner of the building. The crane lifts the first plank for that corner into position. Workers may adjust the plank into precise position with tools such as crowbars. The crane then installs the adjacent plank in the same way. The tongue-and-groove or shiplaps along the edges of the panels are fitted together. The other planks follow in sequence.

Openings are formed with special planks that have cutouts or notches to create the required open space.

When all planks are in their correct positions, it is important to connect them permanently. In some cases, the planks have embedded weld or bolt plates. In this case they are welded or bolted to one another and to the walls. In other cases the planks are sized so that they do not extend to the outer surface of the walls below. In this case, the crew sets rim forms around the perimeter of the building and casts a bond beam around the panels. For the bond beam, steel reinforcing bar is set in the beam area on chairs.

In some projects, a flatwork crew casts a continuous layer of concrete 1 inch to 2 inches thick over the top of the planks. This may be done to create an especially smooth floor surface. It may also be done to create a connection between the floor and the walls, as discussed under "Connections." A concrete topping may be installed later. When a bond beam is cast, the topping may be placed at the same time as the beam.

Pitched roof

When planks are used to create a pitched roof, there will usually be a steel ridge beam in place. The top ends of the planks rest on the beam. It is critical to secure the planks in position immediately so that they do not slide. This may be done with bolts embedded in the planks. These match up with other bolts on the walls and ridge beam. A bolt plate fits over the matched bolts and nuts are used to tighten the plates down, connecting the planks to the wall or beam.

Connections

The connection of a deck to the walls is usually by one of three distinct methods.

The first is welding. The planks have weld plates embedded along their edges where they meet the walls. The walls have corresponding plates in a matching position. Once all floor and wall panels are in place, welders joint the plates for a permanent connection.

The second method is similar, but uses bolts instead. The floor planks and wall panels have embedded bolts that line up with one another along the floor-wall joint. Plates with holes are fitted over them and bolted securely to bridge the joint

The third method is with a bond beam. This method is most often used when the walls include cast-in-place concrete. The planks are sized so that they bear only 3 to 4 inches on the concrete walls below. That leaves a gap of a few inches between the outside edge of the planks and the outside surface of the walls. The crew installs rim forms around the outside of the walls, sets rebar in the space, and casts a bond beam around the planks. When the walls are solid concrete or block, the rim forms are normally plywood set against the outside surface of the walls. With insulating concrete form (ICF) or autoclaved, aerated concrete (AAC) walls, the rim forms are usually foam or a flat piece of AAC, which is flush with the outside of the walls. This piece is left in place to continue the outside foam or AAC surface up and around the beam.

With a bond beam, there are normally vertical rebar extending up from the walls below that will extend through the beam and into the walls created above. There may also be *floor connection rebar* that extend out of the walls below. These are bent over so that they lie about an inch above the floor planks. When a topping is cast over the planks, it embeds the floor connection rebar and strengthens the connection to the walls (Figure 16-3).

Stairs may be installed directly to the planks with concrete fasteners. They may also be fastened to a wooden buck that is connected to the planks.

Floor coverings attach as they do onto any concrete floor.

If the ceiling is finished below, this is usually with a plaster product applied directly to the bottom surface of the planks.

Architects

Many architects are familiar with the use of precast panels and will not have difficulty adapting to designing with them. Precast is well documented and the individual manufacturers provide their own materials.

Precast planks provide significant design flexibility. Very long spans are possible without intermediate support. This means that the construction of most decks will be feasible so long as there are adequate walls at each end to support the planks.

Engineers

Engineers are also widely familiar with precast construction. Many have generic software packages that include tools for designing buildings with precast panels. The manufacturers also provide extensive product and engineering infor-

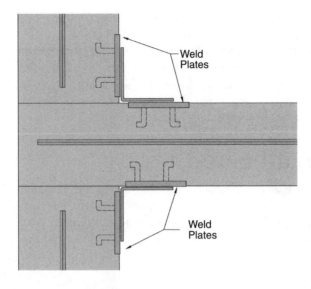

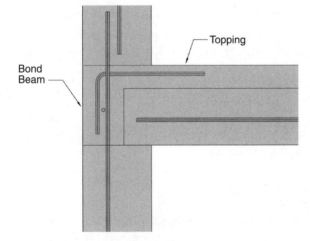

16-3 *Connection of floor to concrete wall with weld plates (above) and a bond beam and topping (below).*

mation. Some have software for their own products available on their web sites.

Specifications of most planks and the recommended rules for using them follow ACI 318. This is familiar to most engineers.

Training

Effective precast installation crews usually require one or two people who are specifically trained in correct installation supervision, sequencing, and procedure. The rest of the crew must be trained in crane procedure and safety measures. However, they may otherwise be lower-skill workers.

Formal training programs for supervisors typically last one day. Many precast manufacturers offer training on their own products. Information on their training is usually available on their web sites. Some local precast associations also offer training. To locate manufacturers or local associations, check the listings of the two major national precast associations. These are the Precast/Prestressed Concrete Institute (PCI) at www.pci.org, and the National Precast Concrete Association (NPCA) at www.precast.org.

Maintenance and repair

There is normally no maintenance or repair required. Like concrete floors in general, precast floors and roofs are resistant to shifting or sagging. They are not expected to need adjustment or repair for the life of the building.

Any finishes installed will wear as normal and may need periodic maintenance or replacement.

Energy

Used as a roof, precast planks provide good air tightness and significant thermal mass. If they were insulated they would provide high overall energy efficiency.

As a floor, insulated planks will effectively separate heating and cooling zones.

In any case, the thermal mass of a floor should help even out and reduce the heating and cooling requirements of the building because of the additional thermal mass it provides inside the building. If a solar heating scheme were used it could provide significant heat storage.

Sustainability

Precast planks will typically help achieve LEED credits for the use of local materials, since they are rarely shipped over 500 miles. Depending on whether they are insulated and how they are used, they may also contribute to the energy efficiency of the building and the award of the related credits. They may contain recycled materials in the form of fly ash in the concrete and recycled steel in the reinforcement. However, to verify this it will be necessary to ask the manufacturer.

Separate from the LEED credits, building with concrete decks is often cited as a method of avoiding the harvesting of trees.

Aesthetics

As noted, any conventional floor covering can be applied to a concrete floor. If the planks are topped with a layer of concrete, the full range of decorative concrete finishes is available to create striking, distinctive floor surfaces. The details of decorative concrete floors are in Chapter 22.

All conventional roof finishes may also be applied. Some roofing is attached with fasteners and is not easily installed in concrete. Instead, furring strips may be attached first, and the roofing fastened to the furring.

If the underside of the floor is finished to create a ceiling, this is normally done with conventional plaster products.

Key considerations

As with all precast installations, it is important to place planks in their proper positions, according to the manufacturers' shop drawings. Errors may not be apparent until there are only a few panels left. At that point, it may be necessary to move many planks to set things right.

The deck must be fully planned before installation. This includes exact dimensions and the position of each opening. It is generally not practical to adjust planks in the field.

Like other precast products, planks may have to be ordered well in advance to allow time for production. The lead time required will vary, depending on how much work the plant has currently.

If the system is used as a roof and the walls are also concrete, the building is likely to be extremely airtight. The heating, ventilation, and air conditioning (HVAC) contractor or engineer should be consulted about measures to maintain proper air intake and humidity control. Possible measures are discussed in Chapters 5 and 12.

When used as a roof, it will be important to use insulated panels. However, once the roof and walls are insulated, the heating and cooling load of the building will likely be sharply lower than in most light frame buildings. Take steps to insure that the HVAC equipment is properly sized for such an energy-efficient building. Chapters 5 and 12 discuss these steps.

Availability

Precast plank is produced in factories across North America. It is important to choose a plant that is reasonably close, as shipping long distances can raise cost. To find a suitable plant, start by searching the databases of the national trade associations representing precasters: PCI, at www.pci.org, and NPCA, at www.precast.org.

Crew availability

Precast installation crews are also available throughout North America. Local precast manufacturers are familiar with area crews or can train new ones.

Support

The manufacturers are set up to provide a wide range of support for the system. They have local representatives in some areas who should be helpful. Otherwise, there is support staff at the plants themselves. For broad questions, the precast trade associations may also be helpful. Contact the PCI through www.pci.org and the NPCA through www.precast.org.

Current projects

The local plants or their local sales representatives should be most helpful in locating projects currently underway that might be visited.

17

Removable form floors and roofs

Removable metal forms can create a concrete floor or roof deck much as they can create walls. In small buildings, removable form decks are most often used on removable form walls. All forms are set together, and walls and decks are cast at the same time. This provides construction efficiencies and connects them into a strong and seamless total concrete shell (Figure 17-1).

Since the forms are removed and reused, a minimum of materials are installed in the floor or roof. The same crews that build walls with removable forms can install the decks. These decks can later be insulated for high-energy efficiency.

Removable forms are particularly efficient in developments with a series of similar buildings. There the forms can be moved a short distance and used many times. This spreads out their cost.

Removable form floors and roofs have grown along with the walls. They have become a realistic and attractive option for creating an all-concrete building. There are currently estimated to be a few thousand houses and other small buildings constructed with removable form walls in the United States each year. Only some of these are constructed with concrete floors and roofs, but the proportion is growing.

History

Floors and roofs cast in place with removable forms appeared with some of the first modern cast-in-place buildings in the late 1800s. Some famous examples are the all-concrete homes constructed with Thomas Edison's new formwork system for houses. These were built in Union, New Jersey, in 1908.

Removable form floors and roofs became widely used in large buildings. However, the technique almost disappeared from small buildings in North America.

17-1 *View of floor forms from underneath.* Western Forms, Inc.

That changed in the late 1990s. This period saw increased demand for concrete buildings. In addition, forming systems and field methods better suited to smaller structures were developed. There followed rapid growth in small buildings with removable form walls. Removable form decks have grown in small buildings ever since.

Many of the forms companies interested in supplying equipment for small buildings were already members of the Concrete Foundations Association (CFA, www.cfawalls.org). As interest in above-ground concrete buildings rose, they organized a subgroup of the CFA called the Concrete Homes Council (CHC, www.concretehomescouncil.org) to advance and promote the use of removable forms concrete systems in this market.

Market

The buildings with removable form floors are mostly in the Southeastern United States, especially Florida, Texas, and such states as Kansas, Missouri, and the Carolinas. The resistance of the construction to high winds is a particular attraction in these areas. Removable form roofs are rarer, and just beginning to be used.

Advantages to the owner

The general advantages of concrete floors and roofs apply to removable form decks, including:

- Resistance of the structure to fire
- Resistance to rot, mold, mildew, insects, and vermin
- Strength and resistance to vibration and settling
- Added structural integrity to the building
- Durability

Floor advantages:

- Reduced sound transmission between floors
- Extra thermal mass inside
- Rigidity and lack of vibration and "bounce"
- Possible creation of a decorative concrete top surface

Roof and ceiling advantages:

- High thermal mass and air tightness for energy efficiency and comfort
- Reduced sound transmission from outdoors
- Resistance to wind damage

It is possible to outfit the roofs with insulation to provide high-energy efficiency.

Advantages for contractors

When building removable form walls, adding decks made with removable forms is relatively easy and economical. The same forms can be used, and they are installed by similar methods. The operation is efficient.

For crews already experienced with walls, deck construction is also easy to learn. If extra or special forms are needed to create the decks, they may sometimes be leased from the manufacturer. It is most efficient to build in the standard dimensions of the form panels. It is also most efficient to use the system in projects where the same forms can be used multiple times in succession.

Although electrical wiring is regularly installed in removable forms concrete, ductwork and plumbing are generally routed so that they do not need to run through the decks.

Components

The decks can be created with the same form panels used for walls. These are typically 8 ft., 9 ft., or 10 ft. tall and 2 ft. or 3 ft. wide. They are flat and smooth on the face that the concrete will be cast against. On the back they typically have a metal frame for rigidity. There are fixtures or fittings to pin or clip or bolt adjacent panels together.

Some forms manufacturers also offer special panels for use in creating decks. These are typically the same design, but with smaller dimensions. This makes them lighter, and therefore easier to lift into place.

Also available are angle and corner panels. Typical corner panels turn 90 degrees. Panels with other angles are available.

The deck assembly

Cast-in-place decks made with removable forms are usually flat slabs of concrete that run horizontally between the walls. Since the walls and decks are typically cast together, they are monolithic. Steel reinforcement runs through the deck and into the walls. This reinforcement is typically in the lower half of the concrete slab. Above this, in the upper half of the slab, is welded wire mesh that stretches across the entire deck.

In cases of very long floor spans, there will be one or more concrete beams below the floor. This typically has a rectangular cross section. The depth of the beam will depend on the strength required of it. Beams are also cast with the floor. Therefore there is no joint or seam between them.

There may be insulation attached to the underside of the deck. Typically this is reserved for roofs. The most common arrangement for this is insulation set between furring strips or studs that are attached to the concrete.

Cost

The total installed cost of uninsulated, removable form floors and flat roofs is approximately $5.00 to $7.00 per square foot. This is the price to the general contractor, and it includes the subcontractor's profit. However, price can vary significantly. It is generally comparable to the cost of other concrete floor systems with similar properties. It will be more expensive than light frame floors on very small buildings, but frame floors do not provide the benefits that concrete does. Cost may be competitive with frame on larger projects.

Table 17-1 gives typical costs for an uninsulated floor on a small building.

Adding studs or straps plus insulation to the deck might increase costs $3.00 to $3.50 per square foot.

Longer spans require thicker decks. This will increase costs of materials. Complex decks raise costs because of the need to assemble special forms for the unusual features.

Costs can drop significantly if there is a series of similar buildings to be constructed one after the other. In this case the same forms can be moved short distances and reused many times. This spreads out their costs.

Costs also tend to be higher in high-wind and seismic areas. High wind and seismic loads tend to require more reinforcement and sometimes thicker concrete.

**Table 17-1. Approximate Costs of Removable
Form Floor Construction on a Small Building.**

Item		Cost per Square Foot*
Labor		$2.00
Materials		$2.05
Concrete	$1.25	
Rebar	$0.45	
Misc.	$0.35	
Pump rental		$0.35
Total		$4.40

*All costs in U.S. dollars per gross square foot floor area.

Code and regulatory

Removable form decks are covered by the general sections on concrete construction in the International Building Code, for the United States, and in the National Building Code, for Canada. Unlike removable form walls, they are not specifically covered in the International Residential Code or the residential sections of the NBC. Therefore most installations require custom engineering by a licensed engineer.

Reinforced concrete cast with removable forms has a track record of over one hundred years. It is familiar to virtually every building official. With accompanying engineering it is generally readily accepted by local building departments.

Installation

Crews on very small jobs like houses are typically three or four workers. These might include a skilled supervisor, a semi-skilled assistant, and one or two laborers. On a larger project there may be multiple teams of four or five workers, each working on a different part of the building.

Uninsulated floor

When the wall forms are in place, the job of setting the deck forms is clear. The crew needs to erect a set of horizontal forms that extend across the entire interior space.

Before setting forms, the crew installs lines of shoring. This will hold the forms up through their assembly and until the concrete cures. Typically there is one line of shoring about every 5 ft. to 6 ft., plus one line close to each wall.

Setting the deck forms begins in the corners. They may be butted to the wall forms three different ways. One is simply to set them on top of the wall forms and connect them. The shoring supports them farther away from the walls. In a second method, the deck forms butt up against the side of the wall forms. In the third method, the crew sets 90-degree corner forms sideways on top of the wall forms and bolts the two together. This creates the right-angle turn from the wall to the deck. Flat deck forms will then butt up to the edge of the corner forms and connect to them.

The crew sets forms all along the walls. They then work inward, setting more forms against the last row and connecting them. Ideally the building dimensions are sized so that the forms fit precisely when they meet at the center. If instead there are gaps that standard forms will not fit, the gaps can be filled with temporary forms constructed of lumber.

When a complete deck of forms is installed, the crew can insert forms for openings. These are usually wooden frames set on the deck forms to prevent the flow of concrete into the area of the openings. The frames may be temporary blockouts, to be removed after concrete has cured. They may also be permanent bucks that remain in place. Bucks are typically outfitted with anchor bolts. The bolts are connected to the buck lumber. Their bent ends extend into the deck space to embed in the concrete and secure the buck to the deck.

The crew also installs reinforcement. This consists of structural reinforcement and mesh set on chairs. Usually the chairs are selected to position the structural bars in the lower half of the slab. The ends of the bars are normally bent to extend down into the walls for a distance of about two feet. The spacing and size of the bars depends on the floor strength required (Figure 17-2).

17-2 *Forms and reinforcement in position for a floor.* Wall-Ties and Forms, Inc.

Mesh is installed for crack control. It is normally set on higher chairs to position it in the upper portion of the slab.

Any conduit for electrical wiring to ceiling fixtures is also installed before concrete placement. It generally extends to an interior wall or connection point for easy wire pulling.

Concrete is placed in the walls and then on the deck in one continuous operation. By directing the line from the pump, the workers deposit the concrete along the top of the deck forms. Some of them rake it into position. One usually draws a vibrator through the concrete to consolidate it. Crew members use tools to level and smooth the surface. When the concrete reaches adequate strength, the crew can strike the shoring and unbolt and remove the forms.

Beams

Beams are formed below the slab with right angle forms. These are connected to the deck forms to turn downward and create a long, rectangular space below the deck that extends from one sidewall to the other. Reinforcement is set in the beam pocket. The beam pocket is filled along with the rest of the deck.

Roofs

A flat roof is constructed like a floor. There is only one significant difference. In the case of a floor, there will likely be structural reinforcement in the walls that extends up past the top of the floor slab. This rebar will extend into the upper-story walls above. In the case of a roof, the wall reinforcement will not extend beyond the top of the slab.

Pitched roofs require placement of the forms at a slope. The lines of shoring are set to follow the height of the roof up to the ridge (Figure 17-3).

The connection of the roof forms to the wall forms may be handled in any of the three ways this is done for flat decks. If an angle form is used, it would usually be a special form with an angle that matches the roof pitch. Forms are also available to create the angle at the ridge of the roof. It is also possible to construct wooden forms to use instead of the special-purpose metal forms, if that is desirable.

The concrete is placed on a slope. This is fairly readily done on pitches of up to 4-in-12. However, many crews find it difficult on steeper roofs. They use a low-slump concrete and push upward as they finish it to counteract its tendency to slide down.

Insulation

Insulating decks is typically done after they are complete and the forms are stripped. This is normally reserved for roofs, since insulation is of less value

17-3 *Forms for second floor and pitched roof.* Western Forms, Inc.

in most floors. The usual method of insulation begins with attaching studs or furring strips to the underside of the deck with concrete fasteners. Then the insulation goes between wood members just as it would go into the bays of a stud wall. Outfitting a deck with this arrangement also provides a plenum for running ductwork, pipes, and wiring.

Connections

The decks are automatically connected to the walls with the reinforcement extending between them and the concrete cast in one pour (Figure 17-4).

Stairs are installed in an opening made for them. They fasten directly to the slab with concrete fasteners, or they fasten to a permanent wooden buck. Some form systems even have forms for casting stairs out of reinforced concrete with the walls and deck.

Floor coverings attach as they do onto any concrete floor.

If the ceiling is finished below, this is usually with a plaster product applied directly to the bottom surface of the concrete. If the bottom of a deck is furred or studded, it may be finished with conventional gypsum wallboard attached to the wood.

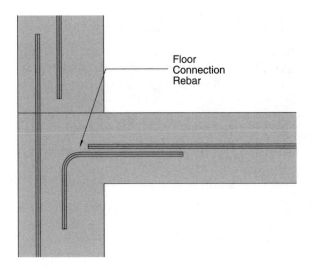

Floor
Connection
Rebar

17-4 *Connection of a floor to removable form walls.*

Architects

Architects are generally familiar with cast-in-place concrete and do not have trouble designing with it. It may be helpful to familiarize them with the types of forms the contractors will use so that they choose dimensions and building features that can be readily created with these forms. This avoids the need and expense of renting special forms or building temporary forms as fill-ins.

Engineers

Engineers are also widely familiar with cast-in-place construction. It is simply a form of reinforced concrete, which almost all structural engineers learn about in detail in their training. The rules of reinforced concrete allowed by the buildings codes are in ACI 318, which is also familiar. For some engineers, using reinforced concrete in small buildings will be new, and may require them to adjust their thinking.

Training

Most crews that build with traditional removable wall forms can adapt easily to using the same forms to create decks. Usually formal training is not necessary. The literature and other materials from the forms supplier will show them what they need to know to connect their forms in these new ways.

It is not usually recommended to have the work done by a crew that is not experienced with traditional forms. However, if this is necessary, at least the lead person should be trained. Some trade schools provide this training. Most forms suppliers offer some form of instruction for new contractors.

Maintenance and repair

There is normally no maintenance or repair required. Like concrete floors in general, removable form floors and roofs are resistant to shifting or sagging. They are not expected to need adjustment or repair for the life of the building.

Any finishes installed will wear as they normally do, and may need periodic maintenance or replacement.

It is important to remember that narrow cracks in concrete are common and rarely compromise the structural integrity of a deck. Most appear soon after casting and should be filled before painting or coating. If any wide cracks appear, it is wise to have an expert look at them.

Energy

Used as a roof, a removable form deck should provide very good air tightness and significant thermal mass. It should also be insulated to be energy efficient. However, with good insulation added, its energy efficiency will be high.

The thermal mass of a floor should help even out and reduce the heating and cooling requirements of the building because of the additional thermal mass it provides inside the building. If a solar heating scheme were used it could provide significant heat storage.

Sustainability

The decks will typically help achieve LEED credits for the use of local materials, since the concrete and steel are rarely shipped over 500 miles. Depending on whether they are insulated and how they are used, they may also contribute to the energy efficiency of the building and the awarding of the related credits. They may contain recycled materials in the form of fly ash in the concrete and recycled steel in the reinforcement. However, to verify this, it will be necessary to ask the manufacturer.

Separate from the LEED credits, building with concrete decks is often cited as a method of avoiding the harvesting of trees.

Aesthetics

As noted, any conventional floor covering can be attached to this or almost any other concrete floor. In addition, the full range of decorative concrete finishes is available to create striking, distinctive floor surfaces that would be impractical on floors not made with concrete. The details of decorative concrete floors are in Chapter 22.

All conventional roof finishes may also be applied. However, roofing that is normally attached with fasteners is usually not fastened directly to the concrete. Instead, furring strips may be attached first and the roofing fastened to the furring.

If the underside of the floor is finished to create a ceiling, this is normally done with conventional plaster products. Wallboard may be fastened to studding or furring strips, if they have been installed. Installing a suspended ceiling is also possible.

Key considerations

Removable forms construction is most efficient if the buildings can be designed and dimensioned to use the available forms without custom fill-in pieces. It is also more efficient in projects with repeated elements, such as developments with similarities between the houses.

If both the exterior walls and the roof of the building are concrete, it is likely to be extremely airtight. The heating, ventilation, and air conditioning (HVAC) contractor or engineer should be consulted about measures to maintain proper air intake and humidity control.

When used for a roof, it will be important to insulate. However, once the roof and walls are insulated, the heating and cooling load of the building will likely be sharply lower than in most light frame buildings. Take steps to insure that the HVAC equipment is properly sized for such an energy-efficient building. Chapters 5 and 13 discuss these steps.

Availability

Several major forms suppliers have systems that are well configured for deck construction. They sell their products across North America through various distributors. Most of them are members of the Concrete Homes Council (CHC) of the Concrete Foundations Association (CFA). They can be found by contacting the Council (www.concretehomescouncil.org) or the association (www.cfawalls.com).

Crew availability

There are thousands of concrete forming crews across North America. Not all are experienced in constructing decks. However, this is a relatively easy adjustment that their forms suppliers can help with. The local distributors of forms can often help locate an experienced crew or bring a new one up to speed. The forms suppliers can provide contact information for their local distributors. Contact them through the CFA.

Support

Most necessary support for removable form deck projects is available from the forms manufacturers that market their products for this application. Staff at their headquarters or their local distributors may be the most appropriate source, depending on the question. For general issues, the CFA may be helpful.

Current projects

Local distributors of the forms may know which local contractors use their equipment to construct decks. This can lead you to projects that you can visit.

18

Autoclaved, aerated concrete floors and roofs

Long, reinforced planks of autoclaved, aerated concrete (AAC) are available for constructing decks. As with precast planks, the AAC units are lifted with a crane and fixed into position. With the workable AAC materials the planks can be modified to some degree in the field. The resulting deck has good insulating properties (Figure 18-1).

Use of AAC decks is fairly common on buildings with walls constructed of AAC panels. The crane and other equipment and materials are already on the job site and can be used to install the decks.

Autoclaved, aerated concrete decks came to North America a little after the walls, and are finding growing use here. Probably a few million square feet of the deck plank is sold for use in North America each year. In the late 1990s, the manufacturers formed the Autoclaved Aerated Concrete Products Association to represent them and advance the use of the materials.

History

Deck plank made of AAC has been manufactured and used in Europe for decades. However, when AAC sales began in the United States in the mid-1990s, the initial projects with AAC walls had decks constructed with conventional materials. By 2000, the deck plank was a regularly offered item of U.S. manufacturers. It was sold as part of the full package of AAC materials for constructing a building. It has gained sales continuously since then.

18-1 *Installing autoclaved, aerated concrete floor plank.* AERCON Florida, LLC

Market

The pattern of sales of the deck plank follows that of AAC in general. About 75 percent goes to commercial and industrial buildings and the rest to residential. The uses are concentrated in buildings that house people, or require a precisely controlled environment for other reasons. The nonresidential sales include hotels and motels, schools, libraries, and certain warehouse and manufacturing facilities. The residential sales include multifamily buildings and high-end homes. The bulk of sales are for projects in the Southeast and the Southwest, where the manufacturing plants are.

Advantages to the owner

The general advantages of concrete decks apply to AAC decks, including:
- Resistance of the structure to fire
- Resistance to rot, mold, mildew, insects, and vermin

- Strength and resistance to vibration and settling
- Added structural integrity to the building
- Durability

Floor advantages:

- Reduced sound transmission between floors
- Extra thermal mass inside
- Rigidity and lack of vibration and "bounce"
- Possible creation of a decorative concrete top surface

Roof advantages:

- High thermal mass and air tightness for energy efficiency and comfort
- Reduced sound transmission from outdoors
- Resistance to wind damage

AAC planks are also good insulators. This makes them energy efficient as roofs. As floors they are also efficient for dividing stories of the building into different heating and cooling zones.

Advantages for the contractor

As with precast planks, installation of the AAC goes quickly. Typically a half-day is enough for the floor or roof of a smaller building. Because installation involves a limited amount of wet materials, it can proceed in a wide range of weather conditions. Few special tools are required, and only one or two crew members need to be highly skilled.

Because AAC can be cut with common wood saws, it is practical to make some modifications to the planks in the field. This must take into account that there is also steel reinforcement in the units.

A crane is necessary for installation. AAC planks are relatively light, and may be lifted with smaller cranes.

The planks must be ordered in advance. Therefore the design and layout of the deck needs to be worked out early.

Components

AAC planks are typically cut to length in the factory, up to 20 feet. They are 2 feet wide, and 8, 10, or 12 inches thick. The thicker panels have more strength to handle greater loads and longer spans.

Each plank has a notch along its edge corners. When two planks are set side-by-side this creates a groove along their joint.

Steel reinforcing bars are embedded in them to provide the needed tensile strength. The bars extend the length of the planks. The planks are made with different sizes and numbers of bars, depending on the structural requirements.

Conventional rebar and concrete for topping are also used.

The deck assembly

The planks are arranged side by side. Their ends rest on the walls below, with a bearing of about 4 inches. Along each joint between two planks is a groove in the AAC where the notches on the edges of the planks match (Figure 18-2).

The deck is topped with a layer of concrete 1 to 2 inches thick. This concrete also fills the grooves along the panel joints. Embedded in the concrete in each groove is one reinforcing bar, which runs the length of the deck and extends into the walls at each end. Around the planks is a reinforced concrete bond beam. The bond beam typically contains two lines of rebar running around the entire perimeter of the building.

If there is an upper floor above this deck, there will be vertical rebar in the lower walls that extend through the bond beam and protrude into the air. The rebar will be embedded in the upper walls when those are built.

Cost

Subcontractors typically charge $8.00 to $10.00 per square foot for installed floors using the system. The square footage includes any openings in the floor. The price includes all materials, all labor, and the subcontractor's markup. This is comparable to other concrete floors that include insulation. In a small building the cost will typically be more than light frame floors, but it will have the advantages of concrete.

Estimated costs for a typical floor project are listed in Table 18-1.

Costs vary with the size and geographic location of the project. The cost per square foot tends to be lower on larger projects because the setup is spread over a larger floor area. Location is important. Projects far from a manufacturing plant result in higher charges for the planks to cover the longer shipping. Complex floors with many openings, corners, and odd angles are more ex-

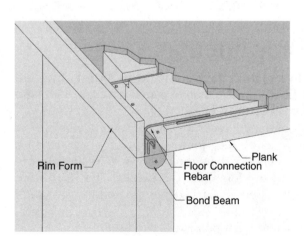

Rim Form

Plank

Floor Connection Rebar

Bond Beam

18-2 *Cutaway view of an autoclaved, aerated concrete floor.*

**Table 18-1. Approximate Costs of Autoclaved,
Aerated Concrete Floor Construction.**

Item		Cost per Square Foot*
Labor		$2.00
Materials		$4.00
Planks	$3.50	
Concrete	$0.20	
Misc.	$0.30	
Crane rental		$0.35
Pump rental		$0.20
Total		$6.55

*All costs in U.S. dollars per gross square foot floor area.

pensive because of special panels, cutting, and the need to keep track of the details. Also important is the span of the floor. Longer spans require deeper planks and thicker slabs. These increase materials cost.

Code and regulatory

The Autoclaved Aerated Concrete Products Association (AACPA) is working with the American Concrete Institute (ACI) to incorporate AAC into ACI 318, "Building Code Requirements for Reinforced Concrete." ACI 318 is the document used by U.S. building codes for the rules of design of nonresidential and larger residential buildings. According to plans, AAC will be covered in section 523A "Guide for Using Autoclaved Aerated Concrete Panels" in the next edition of ACI 318.

Currently AAC is not included in the major model building codes: the International Building Code and the International Residential Code (U.S.) or the National Building Code (Canada). An engineer performs custom engineering for each building and presents this to the local building department. The manufacturers can also provide technical information about their products, including evaluation reports. This can help the building departments understand the products.

Installation

AAC decks are usually placed over walls built of AAC panels. However, they can be readily installed over almost any form of concrete wall.

It is important to plan out the design of the deck in advance. Dimensions and openings should be determined as exactly as possible. Some changes to the planks are possible in the field, but generally it is more efficient to order panels to fit.

The manufacturer provides shop drawings that show the location of each plank. Before installation, the tops of the walls may be marked at the plank locations to insure precise placement.

Crews on very small jobs like houses are typically three or four workers. On a larger project there may be four to six workers. Usually one supervisor is skilled and trained in installation. The rest of the crew must be trained in procedure and safety measures. However, they may otherwise be lower-skill workers.

Level deck

Typically installation begins at one corner of the building. The crane lifts the first plank for that corner into position. Workers may adjust the plank into its exact intended position with tools such as crowbars.

Openings are formed with special planks that have cutouts or notches to create the required open space (Figure 18-3).

18-3 *Autoclaved, aerated concrete planks in position for a floor.*

When all planks are in their correct positions, the crew installs one line of steel rebar in each groove. Workers install rim forms around the perimeter of the walls. They install rebar in the beam pocket between the rim forms and the deck planks.

With all of these items in place, workers clean off and wet down the AAC. A flatwork crew casts concrete into the beam pocket and onto the top of the deck. It fills the grooves along the joints and builds up a layer on top of the AAC 1 inch to 2 inches thick. The crew levels out this topping and strikes it smooth.

Pitched roof

When planks are used to create a pitched roof, there will usually be a ridge beam in place on which to place the ends of the panels. It is critical to secure the planks in position immediately so that they do not slide. This is most often done with bolts embedded in the planks. These match up with other bolts on the walls and ridge beam. A bolt plate fits over the matching bolts, and nuts are used to tighten the plates down, connecting the planks and the wall or beam (Figure 18-4).

18-4 *Constructing an autoclaved, aerated concrete roof.* AERCON Florida, LLC

When all planks are in place, rebar, rim forms, a bond beam, and a concrete topping are all installed, just as on a level deck.

Connections

The rebar in the grooves at plank joints typically extend into the bond beam around the perimeter of the deck. The vertical rebar in the walls below are sized to extend into or through the bond beam. All this rebar links the deck to the bond beam and the walls.

Stairs can be fastened directly to the AAC at the openings. Various concrete fasteners can be installed into the AAC to do this. It is also possible to pre-install a buck that is secured with anchor bolts embedded in the concrete topping.

Floor coverings attach as they do onto any concrete floor.

Below, it is possible to rout out chases in the AAC for installation of such lines as piping or wiring. However, rebar may not be cut and the removal of AAC must be limited to avoid compromising the structure. The bottom surface of the deck may be finished with a plaster product applied directly to the AAC.

Specifications

Autoclaved, aerated concrete manufacturers generally follow standards for their planks established by the American Society for Testing and Materials (ASTM). These are included in the document ASTM C 1452, "Reinforced Autoclaved Aerated Concrete Elements—Specification for Physical Requirements for Reinforced AAC Elements."

Some variation in the material is generally permitted so that the manufacturers can tailor their products to specific purposes. Density ranges from about 30 pounds to 60 pounds per cubic foot. The compressive strength may be 300 pounds to 900 pounds per square inch, although it is generally at the high end for deck plank because of the strengths required. The R-value is 0.80 to 1.25 per inch of thickness. The sound transmission class (STC) is about 45 for an 8-inch thick plank.

Architects

Although few North American architects have experience designing with AAC, it is not difficult to adapt to. The design rules are similar to precast plank, which is familiar to many of them. The manufacturers have manuals and documentation that they can provide to assist.

Engineers

Engineers also can adapt readily to the material. The engineering design is similar to that of conventional precast planks. The manufacturers' evaluation reports include engineering guidelines. Their staff can also assist.

Training

One or two trained people are usually adequate to lead a crew in installation. The others should have safety training for work with cranes and heavy materials, but otherwise need not be highly skilled workers.

The AAC manufacturers offer training seminars. These are mostly one-day courses, and should be adequate for the supervisor of a crew. They may also send an experienced person to the job site of a first-time crew. This is highly recommended.

The manufacturers can be reached through the AACPA at www.aacpa.org.

Maintenance and repair

There is normally no maintenance or repair required. Like concrete floors in general, AAC floors are resistant to shifting or sagging. They are not expected to need adjustment or repair for the life of the building.

Any finishes installed on the top and bottom surface will require their usual maintenance and replacement with wear.

Energy

Used as a roof, AAC planks provide high airtightness and good R-value and thermal mass. The R-value typically ranges from 6 to 15, depending on the thickness of the planks and the density of the material. There are virtually no thermal breaks. The thermal mass of AAC is not as high as with conventional concrete. However, there is more of it than there is conventional concrete in other decks because the AAC planks are solid, without cores or beams.

As a floor, the system may be called on to separate heating and cooling zones. No added insulation should be necessary for it to serve this function effectively. The floor should also help even out and reduce the heating and cooling requirements of the building because of the additional thermal mass it provides indoors. If a solar heating scheme were used it could provide significant heat storage.

Sustainability

Autoclaved, aerated concrete decks can increase the energy efficiency of a building. Although this is not usual in North America, the material can be manufactured partially with fly ash, which is a recycled material. Autoclaved, aerated concrete also has an environmental advantage of containing fewer raw materials than conventional concrete.

On the Leadership in Energy and Environmental Design (LEED) scale, AAC decks can help a building qualify for one or more points awarded for energy efficiency. The total points received on this scale will depend on other energy-related parts of the building as well.

Many buildings currently constructed with AAC could also qualify for the point for use of local materials. Most projects are within the 500-mile radius allowed by LEED. Projects far from the Southeast or Southwest might not qualify.

Aesthetics

Any conventional floor covering can be, and routinely is, attached to this or almost any other concrete floor. In addition, the full range of decorative concrete finishes is available to create striking, distinctive floor surfaces that would be more difficult on other types of floors. The details of decorative concrete floors are in Chapter 22.

If the underside of the floor is finished to create a ceiling, this is normally done with a plaster product or conventional wallboard.

Key considerations

As with precast installations, it is important to place planks in their proper positions, according to the manufacturers' shop drawings. Errors may not be apparent until there are only a few panels left. At that point it may be necessary to move many planks to set things right.

The deck should be planned before ordering the planks. This includes dimensions and the position of each opening. It is possible to adjust planks slightly in the field, but that can add to cost. Like precast products, planks may have to be ordered well in advance to allow time for production. If the plant has the necessary types of plank in stock, product will be readily available. Otherwise the lead-time will vary, depending on how much work the plant has currently.

If the system is used as a roof and the walls are also concrete, the building is likely to be extremely airtight. The heating, ventilation, and air conditioning (HVAC) contractor or engineer should be consulted about measures to maintain proper air intake and humidity control. In addition, the heating and cooling load of the building will likely be significantly lower than in most light frame buildings. Take steps to insure that the HVAC equipment is properly sized for such an energy-efficient building. Chapters 5 and 13 discuss these steps.

Availability

Manufacturers will ship the planks almost anywhere. However, shipping costs are higher farther from the factories. Currently there are three factories in the United States, located in Florida and Arizona. The manufacturers may be found in the Directory of the web site of the AACPA.

Crew availability

Most conventional precast installation crews can also install AAC planks. These can be found in local listings. The AAC manufacturers and their resellers may know of crews with experience in particular areas. Both groups are listed in the Directory of the AACPA.

Support

The manufacturers and their regional resellers are set up to provide a wide range of support for the system. Find them on the Directory of the AACPA web site.

Current projects

The best bet to locate current projects to visit is the manufacturers and resellers of the material.

19

Developments in concrete floors and roofs

There are other developments in concrete floors and roofs that can be useful. Some are other deck systems that are new or offer unique capabilities. Others are novel methods or materials that offer some significant advantages in at least some projects.

Block and joist floor

For years there have been floor systems constructed with concrete blocks. In the last five years, these have been refined to become increasingly practical. They have the advantage of being highly efficient in the construction of small decks.

In the modern systems, work begins by placing special welded steel bar joists on the walls. The joists are about 7 inches deep. They are set parallel to one another at intervals of about 16 inches or 24 inches. The ends rest on the front and back walls of the building. The joists have flat bottom chords that extend to either side about an inch (Figure 19-1).

When the joists are in position, workers set concrete blocks between them, resting them on the bottom chord. The blocks used are either 16 inches or 24 inches long to match the spacing of the joists.

With the blocks in place, a small amount of a thin grout is spread over the deck surface. It fills the gaps between rows of blocks. This encases the joists, locking the blocks to the joists to create a reinforced concrete structure. Enough extra grout is deposited to cover the deck with a very thin layer. A worker uses a squeegee to spread this, filling in the rough block

19-1 *Setting blocks on steel joists.* Block Joist® Co., LLC

surface for a smooth deck. After the grout cures, the deck is at full strength. No shoring is necessary.

It is possible to cut holes to run some ductwork, piping, and wiring in the cores of the blocks. If the ceiling underneath is to be finished it can be painted, although this shows the bottom flanges and the rough block. Adding plaster or wallboard can be done by various means.

The system has been used for a wide range of small projects. These include outside recreational decks, high-strength garage floors that support cars above a full basement, and conventional floor decks and roofs.

Light-gauge composite steel joist

The sellers of bar joists for composite steel joist floors have developed a similar joist made with light-gauge steel. It installs much as the original does. The joists are set on the walls and connected with roll bars. Ordinary plywood goes over the roll bars and welded wire mesh is draped over the top chords of the joists. Concrete goes on top. Punch-outs in the webs of the joists allow freedom for running ductwork, piping, and wiring in the plenum (Figure 19-2).

Light-gauge steel joists may make the system even more attractive for small buildings. Depending on the exact gauge required for a particular floor, it may be possible to perform some cutting of the joists in the field. It may also be

19-2 *Light-gauge composite steel joist system.* Speedfloor Holdings Ltd.

possible to fasten wallboard to the underside of the joist directly, without first attaching hat channel.

Two-part roof

A promising new development is the creation of practical pitched roofs that offer most of the benefits of concrete without the costs and logistics of installing concrete on a slope.

Inventive users came up with the idea of installing a flat concrete deck on top of the building and adding a pitched light frame roof on top of that. The concrete deck is usually more than adequate to tie the walls together and bear all the structural loads of a roof. The wood roof on top only needs to withstand wind and snow and shed rain. This requires lower strength, so it may be possible to use less material on the wood roof and reduce costs (Figure 19-3).

Below the deck, the building will have all the benefits of a concrete roof, including safety from disasters, added rigidity to the shell, thermal mass, air tightness, and so on. The attic space would not be as protected, but that space could be used for less critical functions.

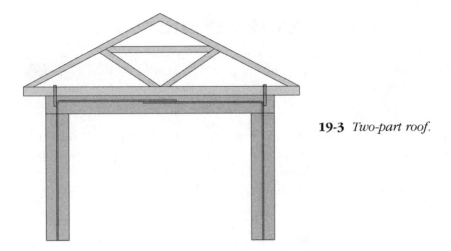

19-3 *Two-part roof.*

A second benefit of this method is that it is not necessary to use an insulated concrete system for the deck. The roof insulation can instead be installed in the frame roof, if that is preferable.

The two-part roof is not a product, but a combination of existing products. It can use any concrete deck system. However, because it is an unfamiliar arrangement, those supplying the frame roof might be uncertain about how to size and design it. Knowledgeable designers and contractors are preferable to avoid unnecessary expense or errors.

Self-consolidating concrete

Cast-in-place concrete decks can benefit from the new self-consolidating concretes (SCCs) much as cast-in-place walls can. Since SCC flows well, it can be placed in one or a few locations on the deck and allowed to move into its final position by itself. Workers raking the concrete over the deck are unnecessary. No one needs to vibrate the concrete, either. All of this saves labor hours. Smaller, less mobile pumping equipment can be used, saving further cost.

The quality of the resulting deck may be higher because of the virtual elimination of any voids.

The pressure that SCC mixes put on the forms may be higher for a time after it is placed because it does not stiffen as much as conventional mixes for the first hour or two. This may require additional shoring or care in placing the concrete for some deck systems. Self-consolidating concrete is also probably not practical for use on pitched roofs. However, as contractors learn how to handle the new mixes, they will likely find significant improvements and cost reductions for level decks.

Fiber reinforcement

Synthetic fibers in the concrete mix are replacing welded wire mesh for crack control in some decks. This may cut the time and cost of cast-in-place deck construction. Several brands of steel and plastic fibers have been formulated exactly for this task. Sold by major companies, they also have long track records and extensive documentation that backs up these uses.

Fibers added to the concrete add cost to it. However, they eliminate the cost of the mesh, the labor of installing it, and the difficulty the mesh causes workers trying to walk on the forms as they place concrete. Depending on the project, these advantages may be significant.

PART IV

EXTERIOR FINISH PRODUCTS

20

Background on exterior finish products

Concrete has excellent properties to serve as the exterior finish of a building. It is durable and requires little maintenance, even when exposed to the elements for decades. It can be inexpensive, and it can be treated with pigments and finish processes to achieve a wide range of attractive appearances. Even installed over a light frame wall, these finishes can give the building some of the advantages of full concrete walls and roofs.

There are now several exterior siding products made with concrete. There is also one practical roofing product. The range of options includes very old, time-proven products and very new, innovative ones. The siding products differ widely in appearance, method of installation, and price. They offer the buyer a wide range of alternatives that can meet almost any set of needs.

The sidings products covered in the following chapters are:
- Stucco (a troweled product)
- Concrete brick (masonry)
- Fiber-cement siding (fastened)
- Manufactured stone (an adhered masonry)

The roof covering is a concrete tile.

History

These different products all have very separate histories. They did not develop together. Most of them are manufactured with different production processes developed at different times. Interest in each of them has arisen separately and often for different reasons. This is unlike the concrete wall systems or concrete deck systems for small buildings. Many of those flourished during the same years because of a general public interest in concrete structures for small buildings. What the exterior finish products do share is a set of properties common to concrete: durability, thermal mass, fire resistance, and so on.

Stucco products containing early cements appear to have existed thousands of years ago. Stucco has gone through countless refinements over the ages, and new, improved formulations are created every year.

All of the other products are newer. Concrete brick did not appear in volume until the development of the concrete masonry machines in the twentieth century. Fiber-cement siding and manufactured stone are even newer inventions, developed along with new manufacturing processes in the 1970s and 1980s. Concrete roof tile is also made in volume concrete masonry machinery but was not widely available until the last couple of decades.

Market

Given their different backgrounds and prices, it is not surprising that the sales patterns of these products are very different. The one thing that is consistent among them is that their sales have all been growing in recent years. It is likely that this is because of the forces described in Chapter 1, which have been driving the sales of nearly all concrete products for small buildings. Those are growing buyer incomes, increased appreciation for the benefits of concrete, and new developments in concrete technology that have helped make the products meet buyer needs.

The market appeal of these products is also quite different. To summarize:
- Fiber-cement siding competes on cost with the most economical siding products, while providing strong advantages over the others.
- Stucco is one of the most economical troweled sidings, but also has the longest track record, good contractor familiarity, and a valued traditional look.
- Concrete brick provides the high quality and durability of any brick with competitive price, more consistent product, and a wider range of colors and styles.
- Manufactured stone offers the accurate appearance of natural stone veneers at a significantly lower cost.

In the roof-covering world, concrete roof tiles provide the advantages of solid roof tiles with competitive price, a very consistent product, and a superior range of colors and styles.

Advantages to the owner

Any exterior finish product needs to protect the rest of the structure from the elements. The concrete products do this while also providing some of the advantages one gets from full concrete walls and roofs. Specifically, these products are durable, fire resistant, have thermal mass, and provide sound reduction. The products contribute these advantages even to light-frame walls, giving them a measure of the benefits of a full concrete wall.

Durability is especially important for exterior finishes. Because of the resistance of concrete to damage, many of these products require no regular maintenance or repair. Those that are painted will require periodic repainting but on a longer schedule than wood products. Typically concrete requires repainting no more often than once every seven years and sometimes less often.

Fire resistance is particularly important in dry areas. Research consistently shows that a resistant exterior finish can protect buildings from igniting from brushfires and wildfires. The amount of thermal mass contributed by these finishes varies widely. However, in some cases is can be almost half as much as the mass in a full concrete wall. Similarly, some of these finishes can contribute almost twenty points to the sound transmission class (STC) of a wall. This is enough to cut the passage of sound energy through the wall by about 90 percent.

Advantages for the contractor

The logistics of installing these products are so different from one to another that each has its own distinct advantages and challenges for the contractor. What they offer as a group is choice. Although they all have advantages of concrete, they come with very different looks at very different price points. There is likely to be at least one product here suited to almost every customer and project.

Components

Most of the components of these finish products are widely different from one product to the other. However, a component that is used in many of these systems is a *weather-resistant barrier*. It is sometimes abbreviated *WRB*, and also goes by the names *weather-resistive barrier* and *water-resistant barrier*. This consists of one or more layers of a thin sheet material such as building paper, building felt, or housewrap. It is used to cover the base wall before adding the finish. Its primary purpose is to prevent water intrusion past the finish and into the structural wall. This is particularly important when the structural wall is light frame material that can rot or rust. Standard practice requires that higher sheets of material lap the lower ones by six inches and side-by-side sheets lap by two inches. The particular material used for the WRB can vary. However, building codes and good practice are generally favoring materials that not only stop the passage of liquid water but still allow water vapor to diffuse through so that moisture will not be trapped inside the wall.

General

Other matters related to these products vary so widely from one to the other that it is best to consider them separately under the discussions of each product in Chapters 21 through 26. This includes such subjects as cost, code status, installation, sustainability issues, aesthetics, and training.

21

Stucco

Stucco is a fine concrete that can be spread in thin coats over a wall surface. It can be troweled onto any common type of wall to create an attractive, durable wall covering. It can also be spread over contours and trim, textured by the installer, and colored or painted to achieve a wide range of features and appearances (Figure 21-1).

The *traditional stucco* product is described here. It is sometimes also called *portland cement stucco*. Most of the volume of traditional stucco is portland cement and sand. It is applied to a thickness of about ¾ inches, creates a hard shell, and allows water vapor to breathe through it. This contrasts with a *textured acrylic finish* (TAF), which is sometimes also called *soft-coat stucco*. This product is a mix containing about half acrylic polymers. It is applied to a thickness of about ¼-inch to ⅜-inch. It is usually installed over a layer of foam sheet sheathing attached to the wall. When it is combined with foam sheathing, it is often called an *exterior insulation and finish system* (EIFS).

Traditional stucco has a track record centuries long and is well understood by designers and contractors. It is durable and reliable, and has generally been free of concerns about trapping water inside the wall. It is one of the more economical wall finishes. Yet the ability to color, shape, and texture it also makes it one of the most versatile.

For centuries, traditional stucco has been one of the leading exterior-wall finishes for small buildings in North America. In the last decade, its popularity has increased as its strengths and its record of reliable performance have come to be increasingly appreciated.

Stucco use has always been concentrated in the homes and small buildings market. In recent years it has been the exterior finish for about 20 percent of these buildings. It is used to cover several billion square feet of wall area in North America each year.

21-1 *Stucco scratch coat.*
Portland Cement Association

History

Spreading a wet, workable material over the surface of a wall (sometimes called *plastering*) is one of the oldest methods of finishing the exterior walls of a shelter. It was quick and easy to do, yet it sealed the joints to protect against the elements. However, the early muds used wore away with time, particularly in rain. With the development of cements, stucco mixes evolved. Adding cement to stucco provided more durability than earlier mixes. Both the Greeks and Romans had a version of stucco containing cement.

In the United States, stucco gained wide acceptance, especially in the southern half of the country. It was a finish of choice on the Federal, Greek, and Gothic Revival style buildings of the 1700s and 1800s that emulated European architectural styles. It also came to be popular in the Northeast and Eastern Canada in Tudor and French style homes.

In 1850, Andrew Jackson Downing advocated the use of stucco in his influential book "The Architecture of Country Houses." In Downing's opinion, stucco was superior to plain brick or stone because it was cheaper, warmer and drier, and could be "agreeably" tinted. As a result of his advice, stuccoed Italian style villas appeared in many parts of North America during the late 1800s.

During the 1900s stucco has remained one of the most popular finishes in the Southern and Western United States and parts of Canada. In Florida, for example, it is by far the most common finish used on small buildings. Its use declined for a time in the Northeast and Upper Midwest. However, in the last ten years, sales have spread back to some of these areas. This appears to be partly because of new stucco formulations that give it great versatility. It is also because of traditional stucco's record of reliability when concerns arose about the durability or moisture performance of some other products.

In 1957, major stucco suppliers and related companies formed the Stucco Manufacturers Association to promote portland cement stucco and educate the public about it.

Market

In the late twentieth century, stucco use was concentrated in the South and West of the United States and particular regions of Canada. However, a 2003 report by the Stucco Manufacturers Association stated, "In recent years, we have seen a marked increase in the use of three-coat stucco beyond the Western residential market." Most of the growth has occurred through ever-greater sales in the Southeast, though stucco has also been spreading to areas farther to the north as well.

Stucco appears to have broad appeal across different segments of the market. As an economical finish, it is common on entry-level housing and lower priced commercial buildings. Yet it is often also selected for higher-end homes and buildings for its appearance or properties. The work is sometimes more intricate in higher-end buildings, but, just as often, it is a straightforward flat finish.

Advantages to the owner

Traditional stucco is durable through all types of weather and environmental conditions. It has high strength and impact resistance. It is readily applied over contoured surfaces, such as rounded corners and moldings.

It may have integral color, which routinely lasts for decades. If it is painted, the paint will endure longer than on wood substrates, so repainting is required less often.

Stucco is widely understood and familiar to contractors, designers, and building officials. It is relatively easy to find a qualified installer, arrange for repairs where necessary, and verify the quality of the work.

Traditional stucco has the ability to breathe, releasing water vapor that might otherwise be trapped in the wall.

Note that the porosity of stucco allows it to retain some moisture. Particularly in wet climates and shady areas of the building, stucco may show some staining over time. Some owners appreciate this "Old World" look. Others will prefer it be painted or cleaned.

Advantages for the contractor

Because of its long history and widespread popularity, plastering is a well-established construction method all around North America and resources are available to assist contractors. Its price is relatively stable through changes in the construction and materials markets.

Stucco can be bonded to any common type of wall. Its flexibility in color, texture, and ability to be installed over a wide variety of shapes makes it appropriate for a broad range of customers and projects.

Components

A stucco installation typically includes two or three different mixes for the different layers. In many cases it also uses a wire lath to reinforce the stucco and hold it to the wall. It generally also involves one or two layers of a weather-resistant barrier set over the substrate to prevent moisture intrusion.

A standard stucco mix is typically one part portland cement to three or four parts sand, by volume. It may also contain small amounts of lime and special ingredients like polymers. Traditional stucco is applied in up to three layers: the *scratch coat*, the *brown coat*, and the *finish coat*. These usually have slightly different proportions of ingredients. The scratch coat has a somewhat lesser proportion of sand than the brown coat. This gives it more cement, and therefore higher strength. This is desirable for the base of the finish. The brown coat typically has more sand. This reduces its shrinkage. The contractor may vary the finish formula significantly to suit the final appearance desired. Typically it has the least amount of sand and finer sand than in the other two coats. This makes it easier to spread to a smooth surface. It may also have integral pigment to provide coloring.

Other admixtures are sometimes added for special purposes. Air-entraining agents increase the workability of the mix, and increase the freeze-thaw durability of the final stucco. Accelerators speed hardening to cut the time required between applying coats and finishes.

Bonding agents help increase adhesion to the substrate, when needed.

The lath may be a mesh of steel wired twisted together in an octagonal pattern, or expanded steel sheet. The weight and spacing of the lath can vary depending on the strength required (Figure 21-2).

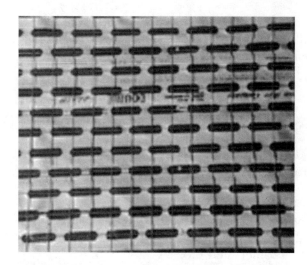

21-2 *Metal lath over building paper.* Portland Cement Association

The finish assembly

The stucco assembly used depends heavily on the substrate it is applied over. It can consist of anywhere from one to three coats, and it may or may not have lath and a weather-resistant barrier. Table 21-1 summarizes the possibilities.

Regardless of the substrate, if the wall has doors and windows or other horizontal surfaces protruding outward from the wall surface, the tops of these are covered with flashing.

The simplest assembly is over concrete surfaces that are essentially flat. In this case, the stucco can be simply one finish coat directly applied to the

Table 21-1. Alternative Stucco Applications.

Substrate	Stucco Assembly Components
Flat concrete surface	Top coat
Average concrete surface	Brown coat
	Top coat
Flat foam	Lath
	Scratch coat
	Brown coat
	Top coat
Light frame wall	Weather-resistant barrier
	Lath
	Scratch coat
	Brown coat
	Top coat

concrete surface. It may be ⅛-inch to ¼-inch thick. Over more typical, slightly uneven concrete surfaces, the stucco consists of two layers. Adhered to the concrete substrate is a brown coat about ⅜ inches thick. It is covered by a topcoat that is about ⅛-inch thick. Over a foam surface, a weather-resistant barrier is often recommended, and lath and all three coats are necessary. The barrier covers the foam, followed by the mesh, covered by a scratch coat about ⅜ inches thick, followed by the brown and finish coats.

A frame wall is generally sheathed with some form of rigid board. Over openings or other horizontal surface is flashing. A weather-resistant barrier completely covers the sheathing. Sheets of the barrier higher on the building lap over the lower sheets and over any flashing. A layer of lath lies over the barrier. It is attached through the sheathing to the studs with corrosion-resistant nails or screws. All three coats of stucco are placed over the lath, with their usual thicknesses.

Large installations require control joints at regular intervals on the wall. These are breaks in the stucco vertically up the wall and horizontally along it. They allow the sections of stucco between the joints to shrink or move slightly without causing random cracks in the field of the section.

Cost

The cost of a stucco finish over wood frame is generally a little less expensive than wooden clapboard or EIFS, and a little more expensive than fiber-cement or vinyl siding. However, stucco applied to a concrete wall surface is typically less expensive than any of these alternatives. This is largely because stucco over concrete requires no barrier, no lath, and fewer stucco coats.

Typical subcontractor quotes for a stucco finish over frame are currently about $5.00 to $6.00 per square foot. Over concrete they are about $1.50 to $2.00. This includes the subcontractor's markup. Table 21-2 contains typical labor and material costs for a stucco finish. These do not include a subcontractor markup.

Materials and labor costs can vary regionally. Both tend to be lower in Southern states and higher in the North. A stucco installation tends to be more expensive over walls with many irregular features because the work is slower.

Code and regulatory

Portland cement stucco is covered in the International Residential Code (IRC). For its requirements, the IRC references two standards. For stucco over concrete, IRC refers to ASTM C 926, "Standard Specification for Application of Portland Cement-Based Plaster." For stucco over concrete, IRC refers to ASTM C1063, "Standard Specification for Installation of Lathing and Furring to Receive Inte-

Table 21-2. Approximate Costs of a Stucco Finish.

Over Wood Frame

Item		Cost per Square Foot*
Labor		$3.35
Materials		$1.10
Stucco	$0.35	
Lath	$0.50	
Barrier	$0.25	
Total		$4.45

Over Concrete

Item		Cost per Square Foot*
Labor		$1.00
Materials		$0.25
Stucco	$0.25	
Total		$1.25

*All costs in U.S. dollars per gross square foot wall area.

rior and Exterior Portland Cement-Based Plaster." These cover such matters as the weight and installation schedule of the lath, ingredients of the stucco material, and the thickness of the application.

When building officials inspect stucco, they most often examine it for material thickness, attachment to the substrate, and any cracks in the surface. Flashing around windows, doors, and other penetrations is a high priority for building inspectors.

Installation

Installation depends on the substrate, since different substrates require different numbers and types of layers. An important requirement regardless of the substrate is proper flashing. Flashing needs to go above all horizontal surfaces extending outward from the wall, such as trim, or door and window frames at the top of an opening.

Over flat concrete

Over a flat concrete surface, a topcoat alone is sufficient. If the concrete is highly absorbent, the installer may wet it first. This prevents it from drawing water from the stucco, which could reduce the water content below what is necessary for good curing. The stucco is usually applied with a trowel. An even coat depends on the skill of the installer.

An alternative method of applying stucco is with a pump machine. These are efficient for a consistent application over a large area. The equipment has some cost, and the efficiency benefits depend on having a skilled operator.

The topcoat is typically installed at the top of the wall first. Then the installer works downward. In this way, any stucco dropping down from above is smoothed out when installation gets down to that level.

The topcoat material may be applied as thin as about ⅛ inch. More may be required over uneven spots in the substrate.

Proper curing of the stucco depends on it remaining moist. Many stucco mixes are formulated to retain water during curing. However, in dry weather special additional steps are recommended. The simplest is to wet down the stucco lightly a couple of times a day for about 3 days. There is now special equipment to create a mist to wet the wall in a process called *fog curing*. Another option is to set up temporary coverings or screens to shade the stucco and slow down drying from wind.

Over uneven concrete

Many concrete and masonry surfaces are rough and have slight uneven spots. Usually two coats of stucco are applied over them, a brown coat and the topcoat. These may be applied by trowel or machine.

The brown coat is typically *floated*, or pressed smooth. This is usually done with a flat wooden tool called a *float*. It increases the stucco density for strength, smooths the surface, and presses it more firmly into the substrate for adhesion (Figure 21-3).

The top coat and curing are done just as with a flat concrete wall.

21-3 *Floating the brown coat.*
Portland Cement Association

Over foam

Some walls are sheathed with a layer of foam. This includes ICF walls, some frame walls, and concrete walls with insulation on the exterior. Since the substrate is softer, the stucco requires reinforcement and three layers for strength.

If the foam is uneven, it is usually leveled out with a rasp or other tool. It is recommended that a weather-resistant barrier then goes over the foam. The lath is next. It must be securely fastened in place. Depending on the type of wall, this may require the fasteners to penetrate the foam and lodge in the structural wall material below. Other foam walls have furring strips or other members near the surface that the fasteners can connect to instead.

As the scratch coat is applied, the installer may *rod* it. This involves running a long metal rod or other straight edge over the surface to cut any high points. This helps insure a flat final surface. After rodding, the scratch coat is scored or roughened. This may be done with a special trowel with teeth. The brown coat will adhere better to the rough surface.

The brown coat, topcoat, and any necessary curing proceed as previously described.

Over light frame

Typically the light frame wall has rigid board sheathing. In this case, installation begins by covering the sheathing with one, or preferably two layers of a weather-resistant barrier. Installation begins along the bottom of the wall so that the upper sheets lap the lower ones. The barrier material must also lap the top of any flashing so that water drains to the outside.

Next the lath is fastened to the studs through the sheathing. The three coats of stucco are then installed, and any necessary curing measures proceed, as previously described.

It is also possible to install stucco over a frame wall without sheathing. Horizontal wire or lath is run over the studs and attached to them. The membrane goes over the wire or lath, and then the stucco lath and stucco are applied. The wire supports the membrane as the installer presses the stucco against the wall. However, some prefer the strength of a fully sheathed wall. Partly for similar strength reasons, this method is not permitted in some areas.

Connections

The key connection is of the stucco to the wall substrate. This is accomplished as noted in previous sections.

Architects

Architects are widely familiar with stucco and have little difficulty working with it. They will commonly specify the color and texture of the stucco to be placed. In large projects, they may require the contractor to prepare sample panels of the material in advance for inspection. The architect will confirm that the samples meet requirements, sign off on them, and compare the final work to the samples if there is any question about the work meeting requirements.

Training

Stucco installers are generally classified as "plasterers," and many do interior-plaster work as well as stucco. Plasterers may learn their trade either informally on the job or formally through an apprenticeship program. Those who learn informally usually begin work as helpers to experienced installers. They are gradually instructed and given responsibility for performing increasing levels of the work. Union plasterers are trained through a more formal apprenticeship program. They work on the job, but also receive classroom instruction. The apprenticeship usually lasts two to three years.

Maintenance and repair

If stucco is painted, it usually requires repainting about every seven years. This compares with about every five years for wood finishes. Stucco can develop cracks over time. These should be repaired to prevent moisture from passing through the finish layer. Periodic inspection to find cracks is advisable. Most cracks are thin. They can be filled with a latex caulk. If they need color to match the rest of the finish they can be painted with acrylic latex.

The occasional larger crack is repaired by first removing loose stucco and blowing out any dust. If the exposed mesh is damaged, it is patched with another piece. The removed stucco is then replaced with a patching compound applied in several thin layers. Patching compounds are widely available from the suppliers of stucco. The finish coat is added after the base has had a few days to cure.

Large cracks may result from shifts in the wall structure beneath the stucco. Although these cracks are not a failure of the stucco itself, they may stem from serious structural problems that should be investigated quickly.

Energy

Stucco may have a small effect on the energy efficiency of a building because it adds a small amount of thermal mass to a wall. By way of comparison, it has approximately $\frac{1}{8}$ as much thermal mass per square foot as a 6-inch thick concrete structural wall. This may even out temperature fluctuations on a light-frame wall slightly.

Sustainability

Since stucco is usually not a large proportion of the total volume of a building's materials, it may not figure heavily into the calculation of a building's sustainability or LEED rating points. It may have some impact on the same dimensions that other concrete products can contribute to: energy efficiency from thermal mass, use of local materials, and (depending on the ingredients of the mix) recycled content.

Aesthetics

The range of possible stucco aesthetics is still growing as the manufacturers continue to develop new variations on the product. In general, stucco installation can be varied in its color, surface texture, and surface patterning. It is also possible to create a wide variety of stucco-covered shapes on the surface by mounting shaped pieces of foam or other suitable materials onto the wall. The installer then applies the stucco over these moldings along with the installation over the rest of the wall.

Key considerations

For a quality stucco application, proper flashing is critical. It is important that all details over openings and at transition points be carefully designed and installed to move water away from the building.

The exact proportions of the stucco mix have a large influence on how easy the material is to put into place. Installers gradually develop a feel for ideal proportioning and come up with their own favorite mixes.

Like any concrete, the final strength and quality of stucco depends on the mix retaining the proper amount of water for an extended time. Adding extra water can result in a weak finish and cracks. If the weather will be below freezing or extremely dry after installation, adequate curing may not occur, and again the result will be a weak finish. Stucco should be applied only when the temperature will be above freezing for some days afterward. In dry and hot regions, special curing measures to keep the stucco moist are likely to be especially important.

It is also important to maintain the full thickness of the stucco finish. It can be tempting to thin the application to save on time, effort, or money. However, thinner sections can be less durable and water-resistant.

Availability

Stucco, lath, and related products and accessories are available from plaster supply and retail stores across North America.

Crew availability

Plasterers qualified to do stucco work are also available across North America. They can be found in any comprehensive listing of local contractors or businesses under such headings as "Plasterers" and "Stucco contractors".

Support

Qualified installers typically turn to their local material suppliers for help in any unusual situations. For certain broad questions, it might be helpful to contact the Stucco Manufacturers' Association (www.stuccomfgassoc.com)

Current projects

There will likely be many projects ongoing in most local areas. The best way to locate them is to contact local contractors to ask if they are performing work currently and ask if they would be willing to let others observe.

22

Concrete brick

Brick made of concrete have the same shapes, sizes, and installation procedures as other brick. They offer greater precision and consistency, and a wider range of colors and finishes. They are most often assembled into a conventional double-wythe veneer wall. This has proven over time to be a highly durable exterior wall that sheds water and moisture effectively (Figure 22-1).

Concrete brick is a natural alternative to traditional clay brick. It offers aesthetic and material advantages at virtually the same cost. It is easy to manufacture and widely available. It is often available with a shorter turnaround time. The product has grown steadily as a high-value option for quality buildings.

About one and a half billion concrete brick are produced in the U.S. and Canada each year. This is enough to cover approximately 200 million square feet of wall. Estimates are that over 80 percent of these brick, or over 150 million square feet, went to covering the walls of small buildings.

History

Concrete brick was a relatively easy product to develop. It is a simple shape that can be produced on a conventional concrete block machine with special molds. U.S. block manufacturers began to experiment with it and released a concrete brick for sale in the 1960s. Use gradually grew as buyers learned to appreciate the product advantages.

By the 1980s, volumes became large enough that it was possible, in some places, to run a machine nearly full time making nothing but brick. At this point producers began to adopt special machinery designed in Europe to produce a large number of small, low-profile units like bricks or pavers instead of a smaller number of blocks. Efficiencies rose and costs declined.

Concrete brick's popularity often got a boost during construction boom times. In these periods, the lead-time for ordering traditional brick often climbs, while the concrete product can be available more quickly. In the recent 2004–2005 expansion, buyers have reported lead times of up to six

22-1 *Worker setting a concrete brick veneer.* Portland Cement Association

months to get the color and style of brick specified for a project. In contrast, runs of almost any style of concrete brick can be available within one month.

Market

About 80 percent of sales are estimated to be for residential construction, around 15 percent for commercial, institutional, and industrial buildings, and the remaining 5 percent for miscellaneous and non-wall applications.

In most parts of the construction market a full brick veneer is most often used on higher-quality construction projects. Partial brick veneers are more often used in the middle of the market. These usually consist of a brick wainscot that covers the lowest several feet of the walls.

Currently, the highest sales are in the Southeast, South Central, and Midwest of the United States and in some locations of Canada. However, brick is used to some degree in all areas of North America.

Advantages to the owner

A brick veneer gives any wall a number of the advantages of concrete. It provides almost a four-inch thick layer of concrete, which is a little over half what most structural concrete walls contain.

A brick veneer adds significant thermal mass and some R-value to the wall for energy efficiency. It increases the wall's sound attenuation sharply. It is highly resistant to fires that may begin outdoors. It is extremely durable. Normal expectations are that it will last over a century with little maintenance.

A brick veneer is generally considered to be a highly attractive, premium finish. Concrete brick provides a wider range of colors and textures than other types of brick.

Advantages for the contractor

With concrete brick the contractor can offer a much wider range of styles and appearances to the buyer. Frequently concrete brick is available on shorter notice. This can reduce project delays.

Concrete bricks are more precise, with more consistently accurate dimensions. This is especially helpful when installing brick in very narrow wall sections, where there is little room to adjust for incorrectly sized units. Concrete bricks are less brittle. There is less breakage and waste on most jobs.

Components

A concrete brick veneer is constructed with the bricks, mortar, and brick ties. Over frame walls a weather-resistant barrier is also often used.

Concrete bricks are available in a wide variety of sizes. Most common is the traditional brick size of 3½ in. wide by 2¼ in. high by 7⅝ in. long. They may be solid or have hollow cores.

Concrete bricks can have any of the finishes and colors available on concrete blocks. These are described in Chapter 6. In general, concrete brick dimensions are accurate to within about one-sixteenth of an inch. However, some finishes, such as tumbling, intentionally make the units irregular to provide an aged or distressed appearance.

Conventional mortar consists of portland cement, small stone aggregate, sand, and hydrated lime.

The wall assembly

Over light frame walls on residential buildings, brick veneers are generally restricted to use over the first story. They are also restricted to a brick width (wall thickness) of no more than 5 inches. These restrictions do not normally apply when the underlying walls are concrete because it is stronger.

Over frame walls

A weather-resistant barrier extends over the entire surface of the frame wall sheathing. Each sheet laps its neighbors, with the upper sheets overlapping the lower ones. Brick ties are fastened through the barrier to the wall studs at regular intervals. The exact interval depends on local requirements related to forces such as wind. A typical spacing is one tie every 24 inches horizontally and every 16 inches vertically. The free end of each tie extends into the brick veneer (Figure 22-2).

The brick wythe is normally set on a *brick ledge*. This is an area at the top of a foundation wall where the wall is widened several inches. It creates a sturdy ledge that extends a few inches beyond the outside surface of the frame

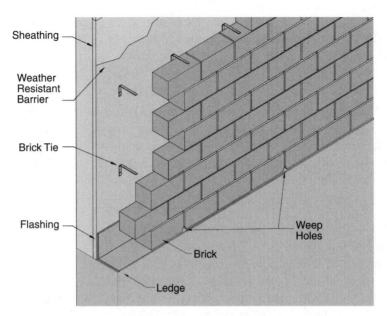

22-2 *Cutaway view of a brick veneer over a wood frame wall.*

wall sheathing. Over the ledge is a flashing that extends up the bottom of the structural wall a few inches, then over the entire ledge and past it slightly. The bricks rest directly on this flashing.

The brick wythe is separate from the frame wall with a cavity in between. The cavity is normally 1 inch to 4½ inches deep.

The bricks are set in mortar. Normally they are arranged in a running bond. However, other bond patterns may be used for decorative purposes.

The horizontal and vertical mortar joints between bricks are normally ⅜ inches thick. However, they can be squeezed or extended to fit the building dimensions. Alternatively, some brick may be cut to size to fit odd dimensions.

The outside ends of the brick ties extend into the mortar in the horizontal joints. They are embedded there.

There are intentional gaps in the vertical joints of the bottom course of brick. These are called *weep holes* or simply *weeps*. Liquid water that may enter the cavity drains out of these holes.

The top of the brick wall normally extends up to the bottom of the roof eaves. If it ends part way up the exterior walls, there is a flashing or some other form of covering over the cavity and veneer.

Over concrete walls

With concrete walls no weather-resistant membrane is required. The brick ties are frequently embedded in the wall. For a block wall their ends are normally embedded in the mortar of the block joints at regular intervals. For other concrete walls they are typically cast into the concrete. As an alternative, they may be attached to the concrete with concrete fasteners.

Insulated wall

The wall may be insulated by adding a layer of foam over the face of the structural wall before installing any brick. The variety and thickness of foam used can be adjusted to achieve the desired cost and R-value. The most common is two inches of extruded polystyrene (XPS). It is generally fixed in place with an adhesive to the structural wall. The brick ties penetrate the foam.

Cost

The price of a brick veneer varies significantly. In the typical one-story residential project in southern states, the subcontractor typically charges $4.00 to $6.50 per square foot. This is the full price including the subcontractor's markup.

Typical labor and material costs for this type of project are summarized in Table 22-1. This includes no subcontractor markup.

Labor rates may be sharply different from region to region. In some northern areas they are close to double the rates in some of the most far southern states. Cost is significantly higher for upper-story walls because of the additional costs of working from scaffolding. The cost will also be higher for more complex walls and lower for very simple ones. Complex walls are those with large numbers of corners, openings, or other irregular features.

Code and regulatory

Concrete brick veneer falls under general masonry veneers because it has a weight of over 15 pounds per square foot. General veneers are covered in Sec-

Table 22-1. Approximate Costs of Concrete Brick Veneer Construction.

Item		Cost per Square Foot*
Labor		$2.50
Materials		$2.30
Bricks	$2.00	
Mortar	$0.20	
Misc. (ties and misc.)	$0.10	
Tools and equipment		$0.20
Total		$5.00

*All costs in U.S. dollars per gross square foot wall area.

tion R703 of the International Residential Code. This requires a cavity between the veneer and the structural wall, with ties between them.

Most major manufacturers also have evaluation reports for their products. These govern some of the finer specifics of installation.

The exterior walls must be adequate to support the veneer. This often limits light frame walls to bearing no more than one story of brick.

Installation

Installation is almost identical to that of other types of brick, which is done by a masonry crew. On a small project such as a house, the crew is typically three or four workers. This includes one or two masons, who set the brick, and one or two laborers, who keep the masons supplied with tools and materials. Other projects may have any number of workers, depending on the size of the job. Typically there is about a one-to-one ratio between masons and laborers.

There must be a brick ledge or other firm structural base for supporting the veneer. The installer places flashing that extends up the base of the structural wall and completely across the surface of the brick ledge. All openings and other relevant transitions must be properly flashed.

Over frame walls

On a frame wall, a weather-resistant barrier such as building paper is installed over the sheathing. It must cover the entire surface that will be faced with the brick. It must also be installed to lap all flashing properly. The sheets at the bottom of the wall are installed first. Their side edges lap one another. The higher sheets are installed later so that they lap over the lower sheets.

Brick ties are fastened to the wall at regular intervals. Flashing goes over the brick ledge.

Installation of the bricks starts at the corners. The mason may erect *storey poles* at each corner to guide the work. These are vertical poles set to the outside corner of the veneer and marked at the top of each course. The corner bricks can be set exactly to the pole and the marks. The bricks between are set to a string line drawn between the poles at the course height.

The first course is set in a bed of mortar on the flashing over the brick ledge. Mortar is also put on the end of each brick to fill the vertical joints as well. However, the mortar is left out of some of these vertical joints on the first course to create weeps.

The second course is constructed on the first in a similar way. The bricks are shifted by one-half unit to create the running bond. The third course shifts back to match the alignment of the first, and this alignment continues to alternate up the wall. Further weeps are created over openings.

Over a concrete wall

No barrier is required over concrete walls. Brick ties are normally embedded in the concrete wall as it is constructed. If they were not, the veneer installer may attach them with concrete fasteners.

Insulated wall

If the wall is insulated with foam board in the cavity, the foam is installed after the flashing and water-resistant barrier and before the bricks. It may be cut in sheets that fit between the rows of ties and adhered to the wall with an adhesive or fasteners. The brick ledge must have been constructed to extend outward by the additional thickness of the foam. This allows sufficient room for the cavity and the brick.

Over a foam-sheathed concrete wall

Insulating concrete forms walls and removable form walls with insulation placed on the exterior have a layer of foam on the side that will back up the brick veneer. They also have embedded plastic strips in the foam, or sometimes furring strips. The brick ties may attach to these strips, or they may need to be anchored in the structural concrete wall to have sufficient strength to hold the bricks. No barrier is necessary. The wall is of course already insulated.

Architects

Most architects are already familiar with brick and how to design with it. Since the rules of designing with concrete brick are the same, they have little if anything to learn.

Some architects not familiar with brick may tend to pick odd wall dimensions that will require extensive cutting of the brick. It is wise to stress the importance of sticking with dimensions in increments that match the brick dimensions. This will control costs and schedule in the field.

Some architects have questions about the product because they are not familiar with brick made of concrete. The manufacturers can usually provide samples and answers.

Engineers

Engineering is usually not required for brick veneers on small buildings. In larger or complex projects it may be necessary to retain an engineer to design such things as a proper brick ledge or select the ties and tie spacing.

Training

Masonry is a skilled trade. Workers usually train as apprentices for about two years before becoming masons. They may do this informally or through a labor union. However, there are thousands of qualified mason contractors across North America prepared to install a brick veneer. Since it is so easy to find a subcontractor to do this work, it is unusual for a general contractor to build or train a brick crew.

Maintenance and repair

One of the major attractions of a brick wall is that it is virtually maintenance-free for decades. Many brick walls hundreds of years old are still standing and in good condition.

Any unacceptable dirt or discoloration of the surface of the brick can usually be removed with a strong solution of granulated soap or detergent and water applied with a bristle brush. Efflorescence is typically cleaned with a bristle brush and clean water. Commercial masonry cleaning solutions are also available for common problems.

After many years the mortar between bricks may sometimes begin to deteriorate in spots. In these situations, it is advisable to *repoint* the brick. This involves scraping out the soft surface mortar from the joints with metal tools and pushing fresh mortar into position to replace it. Masonry crews are normally qualified and available to perform this work.

Energy

A brick veneer adds a significant amount of thermal mass to a wall. The concrete in the veneer is over half the amount in most structural concrete wall systems. It should therefore help to even out the extremes in outdoor temperature and reduce heating and cooling bills.

The brick and the air gap of the cavity add approximately two points to the R-value of a wall. With two inches of foam insulation in the cavity, the assembly adds about twelve points.

Sustainability

A brick veneer can contribute to sustainability by increasing the energy efficiency of the building. It may help to achieve LEED points for energy efficiency.

Like other concrete products, the brick usually comes from the local region and it may have some recycled content. It would be necessary to check with the manufacturer about any recycled ingredients. The veneer may help a project to earn the LEED credit for use of local materials or recycled materials. However, the veneer is only a fraction of total materials use, so its contribution will be limited.

Aesthetics

The broad range of looks available on concrete brick is one of its major selling points. It can be made with any of the colors and finishes of concrete block, which are described in Chapter 6. Some popular finishes are smooth, split-face (rough and irregular), and tumbled (rounded for an aged look) (Figure 22-3).

Concrete brick can simulate the appearance of all popular varieties of clay brick, and looks that are very different and unique.

Key considerations

The brick should be cured for at least a week so that they do not shrink while in the wall. Normally this curing occurs before they are shipped. In rare cases they may be shipped quickly after manufacture, and it would be wise for the user to wait before installing them in the wall.

Availability

There are several major concrete brick plants across North America. They are shipped to all areas. Several producers of concrete brick are listed on the Portland Cement Association web site www.concretehomes.com. Click on "Building Systems," then on "Brick."

Crew availability

Most masonry crews are qualified to install brick. They are included in business directories and construction industry directories of all types.

22-3 *Concrete bricks with a novel rough texture created on the face.* Portland Cement Association

Support

Masons rarely require support for concrete brick work because it is little different from working with other types of brick. When help is necessary, the local brick supplier may be able to provide it. The manufacturers have staff to help with unusual issues or special circumstances.

Current projects

The best possibility to find a current local project is to ask local mason contractors.

23

Fiber-cement siding

Fiber-cement siding provides the look of wood siding with a low cost and high durability. It is a mixture of portland cement and wood fibers that is shaped and textured like lapped wood siding. It is also installed by similar methods. Yet it is less expensive than sawn wood siding, is resistant to water and insects, and needs less painting and maintenance. It is more impact resistant than vinyl siding, and most people consider it to have a more authentic appearance (Figure 23-1).

Because of this attractive set of properties, fiber-cement siding has grown rapidly since its North American introduction a little over 15 years ago. There are now several producers to choose from with dozens of styles available. Since installation is similar to traditional products, there are now many qualified crews, and new crews can learn quickly.

Estimates suggest that fiber-cement siding is now about ten percent of all the exterior wall finish installed on small buildings in the United States and Canada each year.

History

Fiber-cement siding originated in Australia. Its inventor introduced it to North America in 1989. Sales have risen steadily since then. With success of the product, several other companies developed and introduced their own version of the product. The earliest products were made to be lapped horizontally in the style of clapboard, or in boards made to be installed vertically in a board and batten style. But since then, dozens of other varieties have been introduced.

Market

Market penetration is quite different from place to place. In some localities it is now 20 percent to 30 percent of the siding installed in the local market, while in others it has barely been introduced. However, this appears to be simply a

23-1 *Worker installing fiber-cement siding.* CertainTeed Corporation

result of chance and where the manufacturers have focused their promotion so far. Fiber-cement siding is used successfully in all regions and climates.

The product also sells broadly to all levels of the market. As an economical finish product, it is purchased for entry-level homes and modest commercial establishments. Because of its sharp appearance and durability it also sells well in the middle and upper ends of the market. The major factor determining interest appears to be whether the particular buyer wants the look of lapped wood siding. Because of this, sales are concentrated in the traditional lapped wood markets, homes and small multifamily and commercial buildings. They are lower in larger buildings and institutional and industrial buildings.

Advantages to the owner

Fiber-cement siding gives the owner the durability of concrete. Most manufacturers warranty the product for at least 50 years. It is resistant to water, fire, rot, fungus, and termites.

If fiber-cement siding is painted, the typical repainting time is every seven years. If it is purchased pre-primed or pre-painted, the manufacturer typically recommends repainting even less often. It may also be purchased with integral coloring. In this case, no painting is ever necessary. Integral coloring will not fade in ultraviolet light.

Fiber-cement siding does not crack from common impacts as vinyl can.

Advantages for the contractor

The installation of fiber-cement siding is similar to that of other siding products such as wood or vinyl. For this reason, there are few if any new tools the contractor will need for installation. It is also easy to learn.

Often it is supplied pre-primed or pre-painted or even integrally colored. This eliminates steps in the field.

Fiber-cement siding does not cup, as sawn wood may do. It does not crack in cold weather.

The material has a few different properties that the installer must learn for correct handling. However, these are not great.

Components

The siding pieces themselves are the only special component in a fiber-cement siding project. All others are general-purpose items already used on most construction sites. These include such things as nails and caulk.

Fiber-cement siding is made of a mix of cement, fine sand, and cellulose fiber. It is pressed into various shapes and usually given a surface texture like wood. It is cured in a pressurized steam chamber called an autoclave. This provides it with high strength and dimensional stability. The fibers prevent cracking and give it flexural strength.

Most buyers choose siding in a plank shape like sawn wood clapboard. This is available in widths that range from $5\frac{1}{4}$ inches to 12 inches. It is solid material, about $\frac{5}{16}$ inches thick.

Flat planks up to 4 feet by 8 feet are available for assembling into a board and batten pattern. Also available are smaller shapes including shingles, half-rounds, octagons, squares, and others that can be used to cover a building or simply as decorative accents.

The wall assembly

The arrangement of pieces over the wall is largely the same with fiber-cement siding as it is with sawn wood siding. The wall behind it is typically either studs or sheathing installed over studs. Over this should go a layer of building paper or other weather-resistant barrier, although local practice on this varies.

The siding is fastened to the wall with galvanized nails or screws. The fasteners are located so that they attach to the studs. The pieces are lapped the same way as the corresponding wood siding (clapboard, board and batten, shingles, etc.).

Normally the fasteners are attached *blind*. This means that the fasteners are located near the edge of the piece of siding where they will be covered by the next piece of siding that laps over it. This protects the fastener head from exposure and hides it from view. However, in many areas it is necessary to put nails or screws on the exposed parts of the siding to hold it securely. This is called *face nailing*. It would be preferred especially where there are high winds, for example.

Where boards are mated end-to-end, they are simply butted. There is no need for a gap for expansion. Where they mate to corner trim or to window frame or trim, there is a slight gap filled with caulk.

Cost

Subcontractors typically charge $2.25 to $2.75 per square foot for fiber-cement siding installation. This includes all labor and materials, as well as the subcontractor's fees.

The approximate itemized costs of a typical project are in Table 23-1. These include all labor and materials but not the subcontractor's markup.

The costs are fairly consistent across different regions of the country. Labor can differ because of regional variation in wage rates. A more complex wall increases costs because of the extra labor required. A wall is more complex if it has frequent breaks in the siding for such things as corners, openings, changes in elevation, and so on.

Code and regulatory

Fiber-cement siding is still too new to be specifically covered in the major building codes. However, the manufacturers have obtained evaluation reports from the International Code Council (ICC) in the United States and the Canadian Construction Materials Centre (CCMC) in Canada.

Building officials readily accept the product in areas where it has been used. They may need to have more discussion and study the reports if they have not seen it before. They do not usually specifically inspect the siding installation. If they do, they may look for the fastening schedule used and the use of corrosion-resistant fasteners. They may also check to see that the substrate is correctly prepared and the caulking is adequate.

Table 23-1. Approximate Costs of Fiber-Cement Siding Installation.

Item		Cost per Square Foot*
Labor		$1.15
Materials		$0.95
Siding	$0.85	
Miscellaneous	$0.10	
Total		$2.10

*All costs in U.S. dollars per gross square foot wall area.

Installation

Most installations consist of horizontal lapped boards over wood framing. Installation in that situation is described here.

General

Fiber-cement planks and boards may be nailed by hand or nail gun. However, siding staplers do not have adequate power. Staples may also damage the material.

Because of the weight of the product, it must be fastened to the studs and must be attached at least every 24 inches. Therefore only stud walls with spacing every 24 inches or closer will be adequate.

The nails must embed at least 1 inch into the stud. It is important not to overdrive fasteners. This sets the head below the surface, which sharply reduces holding power. It is typically recommended that all nails be at least 1 inch from any edge of the board.

When blind nailing, it is important to use large-headed nails and locate them just above the lap line. Blind nailing is not recommended in high wind areas or on planks wider than 9½ inches.

Procedure

It is usually easiest to install the siding after the trim is in place. Trim is usually conventional wood. However, some new fiber-cement products are becoming available.

Depending on local preferences and requirements, a weather-resistant membrane may be installed over the wall surface first.

Most lapped siding is installed from the bottom up. A starter strip needs to be attached first along the bottom edge of the sided area so that the bottom pieces of siding come out with the same slope as the others.

Each board should be cut to length so it ends on a stud at each end. It must be nailed into each stud. If simple blind nailing is not acceptable, face nailing is used. With face nailing, each nail is positioned along the bottom edge of the board. Because this edge is also where the board laps with the one below it, the nail goes through the top edge of the board below as well. This puts two rows of nails through each board to hold it securely (Figure 23-2).

When one course is done, the installer marks for the bottom edge of the next course to get the desired overlap. The manufacturer may specify a minimum overlap, which is usually about 1¼ inches.

After the boards are in place, the vertical joints are sealed with an exterior grade caulking pressed into the joint. This includes the tight joints between two boards butted end-to-end and the gapped joints where board ends meet trim pieces.

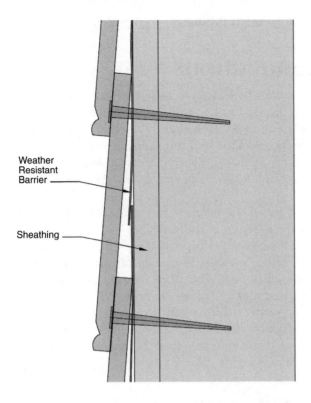

Weather
Resistant
Barrier —

Sheathing —

23-2 *Cutaway view of siding attached to the wall with blind nailing (top) and face nailing (bottom).*

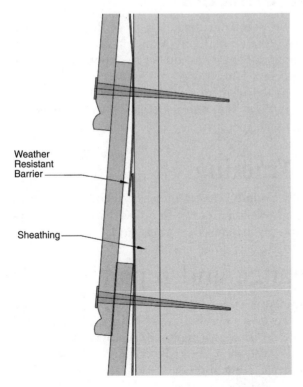

Weather
Resistant
Barrier —

Sheathing —

Connections

Fiber-cement siding may also be attached to most concrete wall systems. To attach to solid concrete walls, typically furring strips are first installed. These are generally vertical, with one about every 16 inches to resemble a stud wall. They are attached with concrete connectors, and the siding is nailed to them. Inch-deep strips provide sufficient depth to drive the nails in one inch, as is normally recommended. However, shallower strips may be acceptable with the right fastener.

Attachment may be different when the exterior surface is covered with a layer of foam. This is normally the case when the walls are built with insulating concrete forms (ICFs) or with removable forms that incorporate a cast-in foam insulation system. In both of these cases, there will normally be plastic strips embedded in the foam for attachment. The strips will typically be vertical, with one every 6 inches, 8 inches, or 12 inches on center. The siding can be nailed to these. Since the strips are shallower than one inch, it will typically be necessary to use a high-friction nail, such as ring shank or galvanized. It may also be important to nail at closer intervals than 24 inches. Consult the siding and wall system manufacturers for specific recommendations.

Architects

Designing with fiber-cement siding is virtually identical to designing with lapped wood siding. The architect must do nothing different. Architects are widely familiar with lapped wood siding and what it can do, so they have little learning to do before using fiber-cement.

Architects may be interested in the aesthetic appearance and the wear properties of the siding. The manufacturers' literature generally provides information that can acquaint architects with these.

Training

Installer training is not normally required. Installation is so similar to that of lapped wood that most experienced siding crews find they can do the work correctly if they simply study the manufacturer's manual first.

Maintenance and repair

The only maintenance normally required is periodic repainting. This will typically be about every 7 years. However, if the siding is pre-primed or pre-painted it can be even less often. If pigmented (integrally colored) siding is used, it need never be painted unless the owner wants to change the color. Fiber-cement siding should be painted with an acrylic paint and never with oil-based paint.

Energy

Fiber-cement siding will add a small amount of thermal mass to the wall and contribute a small increment to the R-value.

Sustainability

The siding may be too small-volume a component of the building to have a significant effect on the overall sustainability of the building materials. However, the siding will in many cases come from less than 500 miles away. This might contribute to achieving a LEED credit for the use of local materials. The cellulose fibers in the siding are sometimes recycled. This could contribute to the amount of recycled content in the building and help receive the LEED point for that property.

Aesthetics

There are fiber-cement sidings available with the accurate look of almost any popular lapped wood product, as well as many other looks. Some of the other surface textures available include very smooth and a stucco-like texture. The stucco surface is created on larger board products.

The pigmented and pre-painted products are now available in a wide range of colors. They are popular not only for their looks, but because they reduce labor in the field and require even less repainting.

Key considerations

The fastening schedule of fiber-cement siding is somewhat different from sawn wood products because of its greater weight. It is important to observe them.

Cutting fiber-cement siding produces silica dust. This should not be inhaled, so use of an inhaler and safety glasses is usually recommended. Check the manufacturer's recommendations. There are now alternative cutting tools that minimize the dust. Some power saw manufacturers offer special blades that produce sharply reduced amounts. Snapper shears and guillotine-type cutters can also be used.

It is generally difficult to hide a wavy wall with fiber-cement siding. A straight wall is therefore important.

Fiber-cement siding should be carried on its edge, not on its face. This is because it tends to bend a great deal if carried flat. This can be awkward or damage the material.

Acrylic paints are recommended. Oil paints should never be used over fiber-cement because the contents of the siding inhibit the curing of the paint.

When it is stored, the siding should be protected from rain. If it becomes slightly wet that is not a concern. However, if it is saturated it can become very flexible, hard to handle, and prone to cracking.

Availability

Fiber-cement siding is widely available at construction products suppliers across North America. The product manufacturers can identify the suppliers that sell their product in any given area. A directory of the major manufacturers is available on the Portland Cement Association web site, www.concretehomes.com. Click on "Building Systems" and then on "Fiber-Cement Siding."

Crew availability

Many siding crews now have experience installing fiber-cement products. Conscientious crews without experience may also do a fine job by studying the manufacturer's installation manual and following it carefully. Many times the local supplier can provide the names of qualified installers.

Support

Generally the local supplier of the product and the manufacturer's literature provides all the support the installer and the owner require. For certain rare or detailed questions it is possible to contact the manufacturers. These companies maintain experts on staff.

Current projects

The best option to find a local project to watch is to contact local siding installers and ask if they have one coming up.

24

Manufactured stone

Manufactured stone consists of highly authentic replicas of natural stone that are adhered to a wall to produce the look of a genuine stone veneer. This is accomplished at a fraction of the cost of a conventional stone veneer. Manufactured stone may be attached to almost any structurally sound wall, new or old. It does not require the sort of structural supporting ledge needed for most conventional veneers (Figure 24-1).

Manufactured stone has become extremely popular over the last few years. The appearance of the product is almost indistinguishable from natural stone. With several major manufacturers, dozens of very different styles are available, simulating a wide range of varieties of natural stone. The ease of installation makes this veneer accessible to a broad range of owners.

There are believed to be tens of millions of square feet of manufactured stone installed throughout the United States and Canada each year. According to press reports, manufactured stone now makes up about 5% of all siding for small buildings. Over twice as much wall area is now covered with manufactured stone compared with natural stone. Its sales are growing an estimated 15–20% per year.

History

The first manufactured stone veneers appeared in the United States in the 1960s. Craftsmen made molds directly from pieces of real stone to get an authentic shape. Over time the production process evolved to produce more and more realistic pieces. Manufacturers began using white cement and hand-applied pigments to replicate the original stones accurately. They adopted lightweight concrete so the manufactured stones would be easier to handle and put less weight on the wall. In the 1980s and 1990s several new companies entered the field. There are now five to ten major producers.

24-1 *Pressing manufactured stones onto a wall.* Tejas Textured Stone

Market

Manufactured stone is installed on small and large buildings alike. Usually it is placed over select parts of the building's walls as a wainscot or accent item. Experts estimate that the majority of product is installed not over 3 feet above ground. However, over some smaller buildings, it covers the entire exterior wall area. An estimated 75 percent of the installation is on commercial and institutional buildings. The remaining 25 percent is residential.

The commercial and institutional projects are predominantly high-traffic buildings where a premium finish and certain atmosphere are desired. Major markets are upscale hotels and motels, restaurants and other retail stores, schools, and libraries. A sizable minority of the product is placed on the interior, such as in a lobby or over a fireplace.

The residential sales are mostly to high-end homes, but also some multifamily buildings. In these buildings the product is almost all used on the exterior.

Advantages to the owner

The major reason most buyers choose manufactured stone is its appearance. It is so close to natural stone that the average person assumes it is natural stone. Varieties are available that mimic dozens of different types of stone. This provides many options for scale, shape, and coloring to match a wide range of design schemes. Yet this finish is available with easier installation and at a fraction of the cost of natural stone.

Because the manufactured stone veneer puts a fairly thick layer of concrete on the building, it also happens to give a wall some of the properties of concrete. Even if the wall is light frame, outfitting it with the product reduces sound transmission, adds thermal mass, and improves fire resistance. Manufactured stone is a durable material, requiring no painting. There may be a need for occasional replacement of broken or detached pieces. Users can save some extra pieces for this purpose from the initial order to assure a good match.

Advantages for the contractor

Manufactured stone allows the contractor to provide buildings with a high level of finish valued by consumers, at a relatively low cost. It is significantly lower weight than a conventional stone veneer, it is easier to apply, and it does not require that the building have a structural support for the veneer such as a brick ledge. The material is much easier to cut than natural stone or normal weight concrete. This makes fitting the veneer into almost any space straightforward. Learning installation is easy for masons and competent workers from other trades as well.

Nonetheless, some care must be taken with the details of the installation. Proper weather-resistant barriers and flashing are important to guiding water away from the wall. This is especially important when the wall is wood frame, as water getting inside may eventually rot the lumber.

Components

The major components of a manufactured stone installation are the stone units themselves. The others are a weather-resistant barrier, common mortar, a wire mesh lath, and corrosion-resistant fasteners.

Manufactured stone pieces are made of a concrete of portland cement, lightweight aggregates, sand, an air entrainment admixture, and mineral oxide pigments. Molds for the pieces come from applying a rubber-type material to actual natural stones and curing it. When it has set, the rubber is peeled away to provide a mold with the authentic shape and surface texturing of the original stone. One product line of manufactured stone is typically modeled after one variety of natural stone, such as river stone, slate, or quarried limestone. The natural stones used to create the molds are selected to provide the desired mix of sizes and shapes (Figure 24-2).

Pigments are generally individually applied to the inside of the molds so that no two pieces come out exactly the same. The concrete is cast into the molds. The pigments absorb into the concrete.

24-2 *Applying mortar to the back of a stone before pressing it into place.*
Eldorado Stone Operations, LLC

The resulting stones have a flat rear surface. Because of the aggregates and the air entrainment in the mix, they are also considerably lighter than they would be if they were made of conventional concrete or they were natural stone.

Dimensions vary considerably even within a product line. Thicknesses typically range from 1 inch to 2½ inches. However, some products designed to resemble dry-stacked stones like ledge stone are made to extend up to several inches out from the wall substrate. The area of a single unit is typically 20 square inches to 720 square inches. The maximum face dimension for most manufacturers is 36 inches. However, producers are constantly introducing new styles and pushing limits on size.

The maximum weight of the veneer is 15 pounds per square foot of face area. This is a stricter limit. According to standard engineering and most building codes, veneers below 15 pounds per square foot can be adhered to a wall and do not need a wall cavity or anchors. This is part of what allows manufactured stone veneers to be so economical.

In addition to the standard pieces with a flat back, there are pieces with an L-angle back to fit over a corner. These provide the appearance of a wall with full-thickness stone.

Some manufacturers offer sections of stones preattached to a backing. This can speed installation by allowing the contractor to attach several stones at once.

The other components are all standard building products. The manufactured stone producer typically has recommended specifications for each of these components. These may vary somewhat from one manufacturer to another and even one product line to another. The recommended weather-resistant barrier (WRB) is usually water-resistant building paper or asphalt-impregnated felt. The mortar is generally a standard masonry mix, or some slight variation on one. The recommended lath is generally either expanded metal or woven wire meeting minimum size, weight, and furring requirements. The nails may be either galvanized or stainless steel. Other fasteners such as staples or screws are also usually allowed. Today, wide-crown corrosion-resistant staples are often used.

The wall assembly

The wall assembly is different depending on what the wall substrate is. As with stucco, there are fewer layers required over a solid concrete wall, including block. Over the entire surface there is a layer of mortar about ½–¾ inches deep that is built up in multiple layers during installation. The flat backs of the stones are set in the mortar. This leaves a face of decorative stones and exposed mortar in the joints.

A layer of lath is added when the wall has a foam surface. The lath is attached with fasteners. If there is a stud wall behind the foam, the fasteners connect through the foam to the studs. If it is a concrete wall, concrete fasteners are attached through the foam to the concrete. The lath is embedded with a scratch coat layer of mortar, and the stones are adhered to the scratch coat.

A conventional stud wall must have sheathing. This can be plywood, oriented strand board, gypsum wallboard (for interior use), or various other products. It also requires the addition of a weather-resistant barrier over the entire surface. Some contractors recommend two layers of WRB material, especially in wet climates. Over the WRB go the lath, scratch coat, embedment mortar, and stones, as with other wall systems.

Cost

The cost of manufactured stone installations varies widely because of the differences in products. Subcontractors typically charge $10 to $25 per square foot of wall area. This includes their labor, all materials, and the subcontractor's markup. This is usually considerably less than natural stone veneers and a bit more than conventional brick veneer. It is more than most other exterior wall finishes, including stucco, fiber-cement siding, vinyl, and sawn wood siding.

The costs of construction for a "typical" project over wood frame are approximately as in Table 24-1. This includes all labor and materials, but not the subcontractor's fee.

Various things can increase or decrease labor costs. Using premium lightweight stones with unusual colors can add $6 to $7 to materials costs. Larger stones are heavier and more difficult to lift, adding labor. Many corners, openings, and irregularities add to the complexity of the job. If the veneer extends

Table 24-1. Approximate Costs of a Typical Manufactured Stone Installation over Wood Frame.

Item		Cost per Square Foot*
Labor		$ 5.00
Materials		$ 6.75
Stone units	$3.50	
Membrane and mesh	$0.60	
Mortar	$0.15	
Miscellaneous	$0.25	
Tools. equipment, supplies		$ 0.40
Total		$12. 15

*All costs in U.S. dollars per gross square foot wall area.

high off the ground, scaffolding may be necessary and work slows. For the most difficult installations labor cost may rise by $4 to $5 per square foot. For the easiest, it might decline $1 to $2. Installation over concrete reduces cost by about $1 to $2 per square foot, because there is no WRB or lath, and there is no labor involved in placing those materials.

Code and regulatory

Manufactured stone-type veneers are accepted by the International Residential Code. Sections R701–703 contain the specific guidelines. Because the product weighs less than 15 pounds per square foot it is classified as an adhered veneer, and no cavity space or mechanical ties to the wall are required.

Most manufacturers have evaluation reports for their products. These provide certain details for how their particular stone is to be installed.

The base wall must be strong enough to carry the load of the veneer. Nearly all common structural walls are adequate.

Building inspectors normally do not perform a detailed inspection of a manufactured stone veneer. When they do they are likely to check such critical details as proper flashing, installation of the weather-resistant barrier, and adequate type, weight and attachment of the lath.

Installation

The installation varies with the type of wall because different substrates require different numbers of layers. The base layers and their method of installation are similar to the application of stucco. In fact, the first coat (called the "scratch coat") of the installation should technically be called a layer of stucco because it is applied to a vertical wall surface. The term "mortar" has come to be used instead because the material is also used to attach pieces of masonry.

It is important that proper flashing and water detailing is completed before the rest of the installation begins. All joints and transitions to other surfaces and materials must be treated in a way that channels water away from the wall.

Installation can be done from the top down or the bottom up. Usually working from the top down is preferable in tall installations because it avoids getting mortar droppings on stones installed below. Working bottom up is often considered easier in low installations (Figure 24-3).

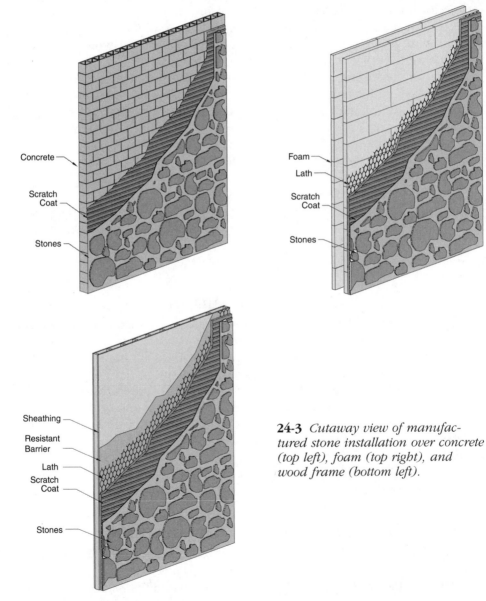

24-3 *Cutaway view of manufactured stone installation over concrete (top left), foam (top right), and wood frame (bottom left).*

Over exposed concrete

The fewest layers are required over a wall of plain concrete or concrete block. The wall must be clean and free of paint or coatings. If necessary, it should be prepared with a cleaner or sandblasting.

The installer applies a first coat of mortar about ½–¾ inches thick. It is given a rough surface with a notched trowel or other tool. This is called the *scratch coat*. The scratch coat is allowed to cure for not less than 48 hours. After this time the installer attaches the stones. Work starts at a corner and progresses toward the center of the wall. This insures that the corner pieces align correctly and any adjustments can be made in the field of the wall. Stones may be cut where necessary to fit. However, cutting can leave a white or uncolored side on the stone that must somehow be hidden.

To attach the stones, the installer has two options. The first is to apply a ½–¾-inch layer of mortar to the back of each stone and press it onto the scratch coat. The mortar should squeeze out around the sides. Alternatively, the installer may put no mortar on the backs of the stones, but cover a two or three square foot section of the wall with a full mortar layer instead. In this case, the stones are pressed directly into the wet mortar. Then work proceeds to the next section of the wall.

When the stones are in place, the installer may tool the joints to create the desired mortar profile. In some cases extra mortar is inserted into the joints with various tools. This may in fact be recommended to help make the assembly more water- and freeze-resistant. It also allows the creation of a fuller style of joint and helps lock the stones into place (Figure 24-4).

Over a foam surface

Several types of walls have an exterior surface of plastic foam. These include insulating concrete forms, removable form walls with the insulation installed over the outside face, and frame walls sheathed with foam. In this case, metal lath is necessary to adhere the mortar layer to the wall.

The installers attach the specified lath to the wall with corrosion-resistant fasteners. Over a frame wall the fasteners must connect to the studs. Typically nails are required to embed at least one inch in the wood. Over concrete walls, it is recommended that the fasteners extend through the foam and anchor into the concrete. This can be done with concrete nails or screws.

After the lath is in place, installation is virtually identical to installation on exposed concrete walls.

Over frame with wood sheathing

When the frame wall is sheathed with conventional plywood or oriented strand board, a weather-resistant barrier must be installed over the wall surface first. The sheets of the WRB are lapped at least 2 inches on their side joints and at

24-4 *Putting extra mortar in the joints.* Tejas Textured Stone

least 6 inches at the horizontal joints, with the higher sheets lapping over the lower ones. Two layers of a WRB material may be recommended for a higher level of water protection.

After the WRB is in place, installation is the same as it is over foam-sheathed frame.

Architects

It is fairly easy for architects to adapt to designing with manufactured stone. The stone can follow the surface of almost any wall without difficulty. Since it is easily cut, there is no hard constraint on the wall dimensions. No brick ledge or other support for the veneer is needed.

The stone manufacturers provide manuals. Many also offer binders including color photos and detailed specifications. Most will send sample pieces to confirm color and texture.

Engineers

Engineers are rarely involved in the design of a manufactured stone veneer. On a larger project, an engineer might check for such items as adequate strength of the wall bearing the stones.

Training

There is no formal training for manufactured stone installers. The manufacturers provide detailed manuals. Because the installation is relatively simple, these tend to be brief and easy to follow. Most installers are experienced masons or stucco plasterers. For these trades, much of the installation procedure is already familiar.

Maintenance and repair

Manufactured stone is generally considered a highly durable product. Some manufacturers provide warranties of up to 50 years. It has the general wear properties of concrete, resistance to fire, insects, and water. Because manufactured stone is an aesthetic product, there may be a need to take steps to maintain its appearance.

Dirt and other deposits may accumulate from time to time. These can be removed with a strong solution of granulated soap or detergent and water, using a brush with soft bristles. Acid-solution products, power washing, sandblasting, and wire brushes should not be used. They may penetrate the pigmented surface and remove the coloring.

In some cases efflorescence may appear on the surface of some stones. This can usually be removed by scrubbing with a brush and clean water.

Occasionally aging can discolor a stone or hard impacts can damage it so that it is desirable to replace it. This is relatively simple. The old stone is cut out and a replacement is attached with fresh mortar. To insure a close match to the original stones, it is a good idea to save spare pieces from the original installation for use as replacements.

Energy

A layer of manufactured stone should add appreciably to the thermal mass of the building's walls. Because it is made of lightweight concrete, it will also have a somewhat higher R-value than conventional concrete. There are no estimates of the effect a manufactured stone veneer on building energy consumption. However, it should provide some reduction.

Sustainability

Like most exterior finishes, manufactured stone's properties are a small influence on the building's sustainability because it makes up a modest part of the whole. It should have a positive effect on energy efficiency. Often it is manufactured close to the job site. It may have some recycled content, although this may be rare. However, any of these attributes would not likely make a major contribution to building sustainability or to the achievement of LEED points because of the small volume of material involved.

Aesthetics

There are now manufactured stone product lines that simulate the look of dozens or hundreds of varieties of natural stone. These include stone varieties of different sizes or ranges of sizes, colors or range of colors, and native to many different areas. Stones from different lines can even be mixed to create novel combinations. There are also special trim and accent pieces to produce looks such as quoins and stone lintels.

Key considerations

Most critical to a quality manufactured stone installation are the water management details. An adequate weather-resistant barrier over wood frame is essential. Some contractors recommend two layers of material, especially in wet climates. Flashings in the wall must protrude adequately to carry the water past the outside edge of the veneer. All flashing and all detailing need to channel water consistently away from the wall and never leave wood exposed.

Availability

Manufactured stone is sold in outlets in every U.S. state and Canadian province. There are several major manufacturers and several more producers of specialty lines. The manufacturers can direct buyers to their distributors in any region. Information on many of the suppliers is on the Portland Cement Association web site, www.concretehomes.org.

Crew availability

Many mason contractors and plasterers install manufactured stone. Numerous crews from various other trades also have experience installing the product. There are few directories with categories for manufactured stone installers. The best options for finding them are to ask local distributors and search general business and construction listings under more general headings like "masonry" and "plastering".

Support

There are many experienced installers, and they generally need little support. When it is needed, the local product distributors can help with most questions.

For unusual or very broad matters, most manufacturers have technical staff that can help. The major manufacturers also provide detailed literature.

Current projects

The best option for finding local projects currently underway is to locate local distributors and installers and ask them.

25

Concrete roof tile

Roof tiles made of concrete are installed as individually lapped pieces, much like slate or clay tiles. They provide a high-quality roof finish to meet any architectural style. They are durable and long lasting. They can significantly reduce cooling costs by providing air circulation, thermal mass, and reflectivity to the top of the building. Compared with other forms of solid roof tiles, they are highly consistent and available in a much wider range of colors, shapes, and styles (Figure 25-1).

Concrete roof tiles are actually a long-established and preferred roofing product in other parts of the world. They became popular in North America during the last fifty years. They are easily installed by roofing crews with slate or clay tile experience. They are manufactured across the United States and in parts of Canada. The advantages and availability of concrete roof tiles have led to steady and continued growth, particularly with owners interested in an architectural, high-performance building.

History

People have covered their buildings with flat, rigid roofing for hundreds of years. A traditional material for the job was slate. Because of its natural layered structure, it was easy to cut or break into the desired flat shape. Yet in place it was very durable.

With the development of fired clay, clay roof tiles became popular. One advantage of the material was that it could be easily shaped in its original, soft state. This made possible pieces that lapped or fit together in new and useful ways.

Soon after modern cement became available, the concrete material was also turned to the job of making roof tiles. The first known concrete tiles were made at the Kroher cement factory in Staudach, Bavaria, in 1844. Some of these tiles are still positioned atop German roofs and are in service today.

Until 1920 they were normally cast as square products that were installed in a diamond pattern. Sometimes color was added to the top, but most were

25-1 *Worker installing concrete roof tiles.* Westile Roofing Products

gray. The first automatic tile machine was developed in Denmark in the 1920s. Called the Ringsted machine, it spread to the rest of Europe, Australia, and South Africa. It made tiles much less expensive through mass production.

In North America, Spanish settlers in the Southeast and Southwest had introduced roof tiles made of other materials. The tiles were popular in these areas. However, concrete tiles were not well known before World War II. After the war the product gained in popularity. Developments in pigments and molding processes added greatly to the range of styles available and the methods of fitting the tiles together. Virtually no other roofing material can produce the architectural variety that concrete tiles can. Sales continued to be strong in Florida, other parts of the Southeast, the Southwest, and the Western states. However, production and sales have gradually spread to more Northern and Eastern regions as well.

In 1971, producers organized the National Tile Roofing Manufacturers Association. It is known today as the Tile Roofing Institute (TRI), and is a nonprofit trade association dedicated to promoting concrete and clay roof tiles.

Market

Currently concrete roof tile covers an estimated 14 percent of the pitched roof buildings in the United States and Canada. This is over two hundred million square feet of roof covered per year.

Sales are primarily in higher-end homes and commercial buildings. Buyers in these segments are more inclined to pay for the greater durability, premium appearance, and lower maintenance of a tile roof.

Tiles are the dominant roof covering in much of the southern half of the United States. This includes much of Florida and parts of Texas, the Rocky Mountain States, and the Southwest. They have also become popular farther up the West Coast and in parts of Canada. More recently they have begun to appear in locations throughout North America.

Advantages for the owner

The most often-cited advantages of concrete roof tile are long-term durability and aesthetics. However, there are others. Manufacturers warrant their products for up to 80 years. Concrete is highly resistant to the elements. The weight of the tiles also resists uplift from wind. Concrete coloring and texturing methods enable the production of tiles that closely simulate other popular materials, as well as tiles with new, unique high-tech designs. The thermal mass of the tiles helps smooth out outdoor temperature extremes, which reduces heating and cooling bills. A tile roof also creates an air gap between the tile and the roof sheathing. Both the thermal mass and ventilation are particularly important in warm climates to reduce air conditioning. The air gap also provides a space where any trapped water can drain and water vapor can circulate away. The sheathing below can remain more consistently dry and last longer.

Advantages for contractors

Concrete roof tile installation is readily adopted by experienced roofing contractors because of its similarity to other types of tile. Yet it provides a range of styles and colors the contractor can offer that is broader than what is available with other products.

Components

Other than the tiles themselves, parts used in the installation of concrete tile roofing are common construction items.

The tiles are available in a large variety of sizes, colors, and profiles. Every year producers release many new ones.

Tiles are categorized by their profile. These are *low, medium,* or *high.* Low profile tiles are about one-half inch thick. They are commonly called *shake, slate, flat* or *bar tiles.* They may interlock along a lapping feature on their side edges. Some are flat on both sides, while some have a weather check or "lip" on the back face along the top edge (Figure 25-2).

Medium profile tiles typically have a profile that resembles the letter "M" or "W." The profile has a ratio of the *raise* to width of less than 1:5. The raise is how far the high point of the tile is from the roof surface below. Low profile tiles lap their neighbor tiles on either side.

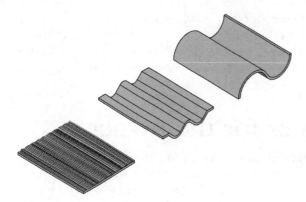

25-2 *Low profile tile (left), medium profile tile (center), and high profile tile (right).*

High profile tiles usually have a profile resembling the letter "S". They have a raise-to-width ratio of more than 1:5.

Most modern roof tiles have nailing holes. There are typically two holes near the top edge of each tile.

The other products used in a tile roof are standard roofing felt, metal flashing, battens, mortar for weather blocking, and plastic roof cement.

The roof assembly

Below the roof tiles is a weather-resistant barrier over the roof sheathing. This is normally a roofing felt. It is in long sheets that extend horizontally across the roof. The top sheet extends over the ridge line on each side and laps the sheet on either side below. These lap the next sheets down, and so on. These are nailed to the sheathing at close intervals along their edges (Figure 25-3).

All along the outer edges of the roof runs a metal drip edge. The drip edge lies over the weather-resistant barrier and extends out slightly beyond the edge of the roof. In all valleys, metal flashing extends from the valley in each direction over the membrane. The drip edge and flashing are nailed into the sheathing along their edges at close intervals.

Roof tiles are set in horizontal rows, or courses, over the barrier and drip edge and flashing. Each course laps the one below it. The bottom course runs horizontally along the roof right over the eaves and extends out past the end of the roof substrate. The course above it laps over it. The third course up laps over the second, and so on. Each course typically has a 3-inch head lap.

At the ridge line the top course on either side does not quite meet. At the ridge line of the roof is a special course of tiles called *ridge tiles*. These tiles are rounded or angled so that they lap over both sides of the roof. Each of these also laps its neighbor on one side. They may be nailed in position or set in mortar.

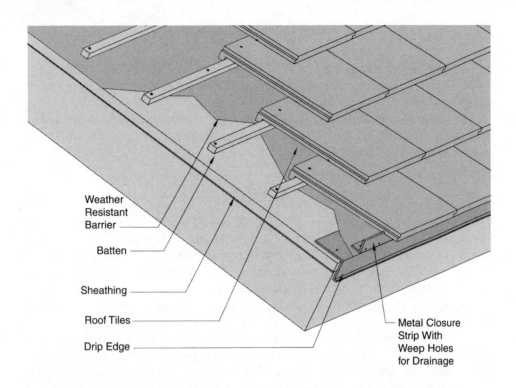

Weather
Resistant
Barrier

Batten

Sheathing

Roof Tiles

Drip Edge

Metal Closure
Strip With
Weep Holes
for Drainage

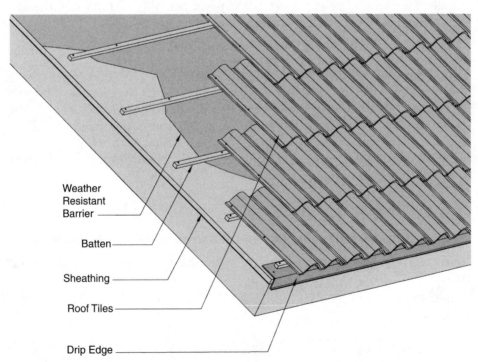

Weather
Resistant
Barrier

Batten

Sheathing

Roof Tiles

Drip Edge

25-3 *Cutaway view of a low-profile (top) and medium-profile roof tile assembly (bottom).*

Some flat tiles have a lip on back of the tile along the top edge. With these tiles, there will be a horizontal batten just beneath the lips on each course of tiles. The lips fit over the batten.

In some installations, there is mortar under the bottom course of tiles over the eaves. This mortar lies under the bottom edges of the tiles all along the eaves, sealing the joint at the outer edge of the roof.

Cost

Subcontractors typically charge about $7.00 to $8.00 per square foot for a concrete roof tile installation. This total charge includes all labor and materials and the subcontractor's markup. The cost is typically less than metal roofing or treated wooden shakes, about the same as clay roof tiles or untreated shakes, and more than basic asphalt roof shingles.

Typical costs of construction for a small roof are approximately as in Table 25-1. These costs do not include the subcontractor's markup.

The cost varies mainly because labor costs do. Labor rates tend to be lower in southern states and higher in northern regions. In addition, labor hours tend to be higher on more complex roofs. Complex roofs are those that have many valleys and connections, changes in pitch or elevation, and so on. The cost of the tiles themselves can vary, too. Some more complex and intricate designs are more expensive than the simplest tiles.

Code and regulatory

Roof tiles are covered in the body of the International Residential Code, the International Building Code, and the National Building Code of Canada. The codes cover such matters as minimum tile specifications, the length of the lap of the tiles, their means of attachment to the structural roof below, the membrane over the roof sheathing, and the required strength of the roof framing to carry the weight of the tiles.

Table 25-1. Approximate Costs of Concrete Roof Tile Installation.

Item		Cost per Square Foot*
Labor		$2.00
Materials		$4.00
Tiles	$3.50	
Membrane	$0.35	
Nails, cement, etc.	$0.15	
Total		$6.00

*All costs in U.S. dollars per gross square foot roof area.

In areas where roof tiles are common, building officials only occasionally inspect the details of the installation. In areas where they are unfamiliar, inspection may be more common and detailed.

Installation
Substrate preparation

A good installation begins with proper covering of the roof sheathing. The crew fastens down a membrane with nails or staples all over the roof. This is typically minimum ASTM type 30 roofing felt. The upper pieces are lapped over the lower, and end joints are also lapped.

At the edges of the roof, drip edge is nailed directly on top of the felt. Standard metal flashing is nailed in the valleys. This is usually 16 inches wide, and is nailed along the edges. Plastic roof cement seals the top edge of the vertical flange and covers all nail penetrations. Flashing and cement also go around all penetrations (pipes, vents, etc.).

The crew then performs the layout directly on the roof. They snap a horizontal chalk line to mark the height of each course of tiles. These are typically every 15½ inches up the roof to give the proper course overlap. If the tile used has a lug on back, the crew installs a line of horizontal batten for each course, just below the line where the lug will fall.

Tile installation

The bottom course of tile, along the eaves of the roof, is installed first. The tiles are set to the bottom chalk line and nailed in position, one by one. Adjacent low-profile tiles are butted side-by-side. Medium- or high-profile tiles are lapped along their ribs. For tiles with a lip, the lip fits over the batten to hold the tile securely on-line. Depending on the roof pitch the tiles may be fastened to the roof with nails or screws through its nailing holes. Tiles may be cut or broken as needed to fit precisely into the existing roof space (Figure 25-4).

The second course goes on in the same way above the first and so on, up to the ridge of the roof. The higher courses then go on in succession. Ridge tiles set over the ridgeline of the roof. These lap one another from one side of the roof to the other. They may be set in adhesive foams or mortar, or nailed through holes under the lap.

Alternative connections

With many tiles, the first course may be set in mortar. This is a traditional technique that seals the edge of the roofing and provides a distinctive appearance. The mortar goes on the edge of the eaves. Each tile is set with its bottom edge

25-4 *Roof tile installation in progress.* Monier Lifetile LLC

in mortar and nailed along its top edge. The installer then presses the mortar with a tool to create a straightedge finish and create a tight seal.

Another alternative connection method is to use adhesives to connect some of the tiles. It may also be adequate to omit some of the nails. However, there are specific schedules for this; these vary with such factors as roof pitch. It is not to be done haphazardly.

Architects

Architects rarely require any special knowledge to design with roof tiles. Roof tiles may be assembled to fit any roof consisting of planar surfaces and almost any roof with irregular or curved surfaces. Roof dimensions do not affect the feasibility of a roof tile installation since the tiles may be cut to fit.

Architects in Southern states, where roof tiles are common, will likely be familiar with them and comfortable designing with them. Architects in other areas may wish to learn more. The major concrete roof tile manufacturers have literature and staff to help them. The Tile Roofing Institute (www.tileroofing.org) also has materials and guidelines helpful to the building professional.

Engineers

The use of concrete roof tiles does not generally require an engineer, since most building codes already address the issue of weight (live and dead loads). Tiles have been designed to generally meet those requirements. In specialty installations such as metal or concrete decks that are considered pre-engineered systems, the design engineer should always be involved. On retrofit or "reroof" projects, the weight of the roofing material may be an issue. In this case there are cost-effective methods to increase the supports, and an engineer can help identify those options.

Training

There is no special or required training for concrete roof tile installation. It is similar to the installation of clay roof tiles or slate roofing. Many roofing crews are familiar with these other products, and installing the concrete is a minor adjustment for them.

Basic training is recommended for at least the lead person on a roofing crew that is unfamiliar with any form of roof tiles. The Tile Roofing Institute offers comprehensive contractor training programs. TRI has also developed a series of installation guides that provides recommendations for moderate, cold, and high-wind applications. For more information, contact TRI (www.tileroofing.org) or the tile manufacturer.

Maintenance and repair

Concrete tile roofs are extremely long-lived. Most manufacturers offer a 50-year life guarantee. If a tile breaks during that time, the replacement may be at the manufacturer's expense. Occasionally one of the other features of the installation, such as a drip edge or mortar, may work loose and require replacement. This is rare, but it is important to replace the damaged part before water can reach the roof below and cause rot or rust.

With concrete roof tiles, the complete roof assembly has a "Class A" fire rating in the building codes. This rating applies to buildings or parts of buildings that are most resistant to damage from fire. They are eligible for some of the lowest fire insurance rates.

Energy

A concrete tile roof adds considerable thermal mass to the top of a building. Although the effects have not been carefully measured, this should reduce heating and cooling costs by reducing the high and low temperature extremes acting on the exterior envelope.

The reduction of cooling needs for other reasons is widely recognized, however. When the hot sun raises the temperature of the tiles, the air space between the tile and the roof below eliminates rapid heat transfer. It also allows the warm air between them to circulate and flow out of the roof, drawing in cooler outside air. In addition, light-colored roof tiles reduce the amount of heat that the roof gains from the sun. Research by the Florida Solar Energy Center suggests that a standard vented tile roof reduces ceiling heat flux by as much as 48 percent, compared to a standard black shingle roof.

Sustainability

The use of a light-colored, self-vented tile roof will increase the energy-efficiency of most buildings. It may help it qualify for at least one Leadership in Energy and Environmental Design (LEED) point for energy efficiency. It may also help to qualify for a credit for use of local materials. The tiles almost always come from within the 500-mile radius allowed by the LEED system. In rare cases the tiles may include some recycled material. It is necessary to check with the manufacturer to determine this for a specific tile.

Aesthetics

As with most concrete products, a wide range of colors is available on concrete roof tiles. It is wider than the range available on clay tiles or natural stone products like slate. Color variation is also possible, and many lines of tile have random gradation in color within each tile or from tile to tile. The tiles are also available with different textures, which can be molded onto the surface. Some of the most popular concrete tiles closely simulate the appearance of clay roof tiles, slate, or wooden shakes. However, many other appearances are also available. This includes looks that would be difficult to achieve with any of these other materials.

Key considerations

Critical to a good tile roofing installation is the correct preparation of the structural roof surface. The correct type and installation of membrane, drip edge, and flashing insure that any stray water does not get to the roof substrate below. Nailing these items down with the proper nailing schedule makes sure they stay down securely.

Availability

The major roof tile manufacturers have as many as thirteen plants across North America, and regional companies serve many areas as well. There are plants and networks of distributors in all regions of the United States and major areas of Canada. It is possible to locate sources for concrete roof tiles on the web site of the Tile Roofing Institute (www.tri.org) by searching its Membership Directory.

Crew availability

Many roofing crews are experienced with installing tile roofs. There are fewer in northern areas, but even there many localities will have numerous qualified crews nearby. When locating one proves difficult, the local roof tile distributors can usually provide references to a few.

Support

Most experienced roofing crews require no support to install concrete roof tile. When they do, they can usually get what they need from the local distributor. For broad questions about the product it may be helpful to contact the manufacturer or the Tile Roofing Institute (www.tri.org).

Current projects

Concrete tile roofing projects are so common that they are not closely tracked by a central source. Local tile distributors may know of particular upcoming jobs or be able to refer to local installers who can give directions to their next installation.

26

Developments in exterior finish products

Architectural block veneer

Architectural concrete block is available in 4-inch wide units to create a non-structural veneer. This is similar to brick veneers, except for the size of the units (Figure 26-1).

Brick and block can also be combined to create a variety of different patterns and appearances. One common practice is to install two to six courses of architectural block at the base of the wall and brick above it. This simulates the traditional masonry arrangement of larger stone units forming the foundation of the building and brick covering it above.

Four-inch architectural block is available from many of the manufacturers of traditional gray concrete block. All the finishes and colors of full-thickness architectural block are available. These are described in Chapter 6.

Installation is by most of the same masonry crews that build full structural block walls and brick veneers. The methods are the same as for installation of a brick veneer. As a result, crews to perform the work are widely available and proper installation methods are widely familiar and well understood.

Cast stone

Cast stone uses careful selection and mixing of aggregates, cements, and pigments to create solid concrete with the look and properties of natural stone. It is shaped into units appropriate for veneers, quoins, window surrounds, balustrades, mantles, and dozens of other items traditionally cut from fine stone

26-1 *House with an architectural concrete block veneer.* National Concrete Masonry Association

26-2 *Masonry veneer with cast stone accents and components.* Continental Cast Stone Manufacturing Inc.

in the buildings of the wealthy. Yet cast stone is a fraction of the cost of its natural stone counterparts.

Cast stone is a solid, normal weight concrete product now produced by several companies across North America. They have mastered the craft of tailoring concrete mixes to mimic many varieties of natural stones and mold them into useful shapes.

The exterior finish products are installed much like other masonry veneers. In fact they can be combined with block, brick, and other products to create the pattern and look desired. Because they fit into the standard masonry construction, they can be readily adopted by the available skilled workforce (Figure 26-2).

Manufacturers produce many cast stone products for use indoors, too. For more information, contact the Cast Stone Institute (www.caststone.org). The web site has a directory of Producer Members that lists manufacturers of the product.

PART V

LANDSCAPE PRODUCTS

27

Background on landscape products

Concrete products are becoming a larger and larger part of the landscape around small buildings. This is partly because concrete's natural advantages outdoors are becoming appreciated. Concrete endures the elements and the wear of traffic. Its weight prevents shifting and keeps outdoor installations stable against forces like ground pressure and wind. For these reasons, it is extremely well suited to outdoor functions (Figure 27-1).

But the *usefulness* of the material only accounts for part of the growth. Today concrete is often installed outdoors because of its new *aesthetics*. When concrete can have interesting colors and textures, landscape architects and property owners find new things to do with it. It not only replaces other materials formerly used for retaining walls and paving. It is also used to create patterns and terraces and platforms that make the landscape more interesting and attractive.

Fundamentally there are only two major landscape functions for which artificial products are commonly used. One is to *pave* the land. *Paving* creates stable, durable walkways, paths, and drives that save traffic from having to move through the mud. The common concrete products for this are *pavers* and *flatwork*. The second function is to terrace the land. This is done with a *retaining wall*, which holds a steep slope in the land. It makes it possible to turn sloping terrain into two or more sections of relatively flat land that are often more usable. Increasingly this function is filled by the *segmental retaining wall*.

However, the practicality and new aesthetics of these products has opened them to new uses. All of them are increasingly employed to create features on the land that are also partly or completely ornamental. This potential for beauty outdoors is part of the benefit of modern concrete.

27-1 *A home outfitted with several concrete landscape products.* National Concrete Masonry Association

History

Natural stones have been used to pave and terrace the land for thousands of years. The Romans were particularly noted for the quality of their roads. The top layer of these roads consisted of flat stones cut or chiseled to shape so that they fit together like tiles. Early retaining walls were simply larger stones piled on top of one another.

When modern concrete was developed in the 1800s, it was a natural for use outdoors. However, the landscape applications were slow to come about. This was partly because the first concrete was expensive. As the cost gradually came down, it became more competitive for these uses that employed large quantities of material. Gradually concrete flatwork became popular. Unlike stones or bricks, it could be cast on the ground in a continuous layer to create walks and drives. Its use has grown steadily. It was not until immediately after World War II that concrete pavers were first developed.

Natural stones found on site and common railroad ties always dominated retaining wall construction because they were cheap and readily available. Some traditional block or removable forms walls were built occasionally when high strength or precision was required. Finally in the 1980s, inventors came up with concrete units that were cost-competitive. They could be stacked rapidly yet accurately. Combining this with the benefits of concrete has led to segmental retaining walls rapidly replacing most of the older materials.

Market

All of these landscape products have come to take a major share of their respective markets. All are still growing. But their properties, their appeal, and the length of time they have been available differ significantly. As a result, their current sales and market shares are sharply different.

All of them appeal to a broad market. The most decorative versions of each product naturally tend to sell more to upscale customers. They are generally specified for higher-end residences and commercial buildings, and less often for institutional or industrial projects. However, each of the concrete landscape products is also available in several very basic, economical versions as well. These versions enable the products to sell well to entry-level housing, lower-scale commercial, and industrial-type buildings for purely functional, non-decorative use.

Advantages for owners

The general advantages of the landscape products to owners are durability and beauty. They have some important potential environmental benefits as well.

Most methods of creating paving and retaining walls function well. The advantage of concrete is that it can remain in good condition and function well for scores or hundreds of years. It is impervious to the common elements, including water, sun, wind, drying, insects, vermin, and impact.

Beauty is a property that has grown rapidly in recent years. The landscape products can be treated with the new methods of pigmenting, shaping, and texturing just as other concrete products are. Most of these effects are far beyond what can be achieved with traditional landscape materials such as asphalt, clay brick, local stone, and treated lumber.

Recently, certain environmental benefits of concrete paving have become prominent. For one, concrete paving can be a much lighter color than conventional asphalt, keeping the area cooler. For another, both flatwork and concrete pavers can be produced in *pervious* versions that allow water to penetrate and drain directly into the soil. This reduces stormwater runoff that can cause erosion and accumulate unwanted materials downstream.

Advantages for contractors

Although these products differ, they all offer a good business opportunity for many contractors. Most of them are easy to learn and are growing. Some are value-added products that command a higher price and profit margin.

Components

The items used with these landscape products are mostly common materials such as stones and sand. One special product used by several of them is *geotextile fabric.* This is a tough plastic sheet material with many holes. It is buried in select locations to stabilize the soil. Stable soil can make installations of landscape products more permanent. However, geotextile fabric is not required in all situations.

Code and regulatory status

The landscape products are generally not tightly controlled by building codes or departments in small projects. Paving and low-height retaining walls have little direct influence on life safety issues. Therefore, with a few exceptions they are not covered in building codes and not closely inspected. Instead, the satisfaction of the customer is the control on methods and materials.

Architects and engineers

On most small projects, the *landscape contractor* does the design and selection of materials. On higher-end and larger projects, there is frequently a *landscape architect* that makes these decisions. Today most good landscape architects are very familiar with concrete products. They have become important tools for creating attractive and usable grounds for a building.

Engineering is rarely necessary for a small landscape project. It is often required for tall retaining walls. However, these are rare as part of a small building project and are not covered here.

Maintenance and repair

One of the major advantages of concrete landscape products is that they typically remain intact, functional, and attractive for the life of the building with little or no maintenance or repair. There may be some need for upkeep of certain decorative surfaces to maintain appearance, as discussed in the appropriate chapters.

Sustainability

Concrete landscape products can contribute to sustainability of a building project in important ways and help qualify for certain Leadership in Energy and Environmental Design (LEED) points.

One environmental contribution possible with concrete paving is that it can reduce unwanted runoff of rainwater. Large volumes of water flowing over the land can cause erosion. They can carry materials and substances to downstream locations where they are unwanted or may accumulate in toxic concentrations.

Both concrete flatwork and pavers are available in *pervious* versions. A pervious material allows water to flow through readily and naturally. A heavy accumulation of water can soak directly into the ground as it naturally would, rather than run over the land. The use of a pervious paving may even contribute to qualifying for a point on the LEED scale. This is LEED credit 6.1, entitled "Stormwater Management." Credit 6.1 lists the scale for projects with environmentally sensitive water management.

Both forms of paving are also available in light colors that naturally reflect sunlight. This keeps the surrounding area cooler than darker paving products like asphalt. In the summer this keeps the surrounding areas cooler. That can increase comfort and decrease air conditioning costs. It might also help qualify for one LEED point awarded to landscaping that mitigates the so-called "heat island effect." This is LEED credit 7.1.

Aesthetics

The landscape products are all relatives of established concrete systems. They can therefore be textured and colored with a very broad range of techniques developed for all of these systems.

Pavers and retaining walls are concrete masonry products manufactured on machinery similar to that used for concrete block. This machinery is described in more detail in Chapter 6. They can also be finished with the techniques commonly used on architectural block.

Concrete flatwork can be treated with all the techniques of decorative concrete. These are described in detail in Chapter 33.

Key considerations

Concrete landscape products have become a rapidly changing area driven by creativity. New products are appearing regularly. Some of these produce new looks and some have further benefits for the environment. Designers are putting even the existing products to new uses. It is impossible to predict all the new products and applications that will become important in the future. Some of those that currently look promising are covered in Chapter 31, Developments in Landscape Products.

28

Concrete pavers

Concrete pavers are small, solid pieces of concrete that are set on the ground to create the surface of a walkway, drive, or road. They make a durable and attractive outdoor surface. The many available colors, shapes, and textures provide a wide range of appearances. Maintenance, when needed, is relatively fast and simple (Figure 28-1).

Pavers have grown rapidly in popularity in North America over the last 20 years. The primary attraction is aesthetics. Any driveway can have the look of the roads of a European estate, a turn-of-the-century mansion, or a celebrity's home. Walkways become as much decorative elements as functional paths. For this reason, buyers in some cases are adding pavers in areas of their lots that they otherwise might have left unpaved.

There are also types of pavers that create pervious pavements. Water can flow through the paving layer and into the ground. This reduces runoff and water pollution, benefiting the environment.

The Interlocking Concrete Pavement Institute estimates that in 2003 over 600 million square feet of pavers were sold in the United States and Canada. Sales grew approximately nine percent from 2002. They have more than doubled since 2000.

History

History's first record of making a pavement with stones dates to 5000 BC with the Minoan civilization. The Romans turned the practice to a highly scientific craft to build their famous roads. They first laid down layers of compacted stone. For the top layer they cut flat stones to shapes that let them fit closely together. Some of these roads are still in use today, 2000 years later.

During the Middle Ages and Renaissance in Europe, cities and the wealthy paved major streets and drives with rounded cobblestones. Later the practice developed of using bricks. These had the advantage that they were flat

28-1 *Paver installation under way.* Interlocking Concrete Pavement Institute

and could be set in close-fitting patterns to make a more nearly continuous surface.

After World War II in Europe, there was a need for rapid reconstruction with durable, readily available materials. The Dutch originated the modern interlocking concrete paver to replace damaged clay brick streets in the late 1940s. The product spread quickly to Germany and the rest of Western Europe for use in both pedestrian and vehicular pavements.

North American concrete block producers began to make and sell the product in small volumes in the 1970s. Pavers could be molded on the existing concrete block machines but not highly efficiently. By the 1990s the many alternative colors, textures, and shapes possible with concrete masonry had been applied to pavers as well. Interest boomed. The manufacturers brought in special machinery from Europe and Japan designed to produce pavers at three times the rate of a standard block machine. In 1993, the Interlocking Concrete Pavement Institute (ICPI) was established. It is a trade association representing the manufacturers and users of concrete pavers in the U.S. and Canada. Most manufacturers are members as well as many suppliers and contractors.

Market

The Interlocking Concrete Pavement Institute estimates that 2003 sales included 530 million square feet in the United States and 85 million square feet in Canada. Some 70 percent of this was sold to residential applications such as patios, walkways and driveways. The residential applications include mostly high-end homes, but the product is beginning to appear in the middle range of this market. The remaining was sold for commercial and municipal uses. These applications include large-scale projects such as municipal plazas, campuses, ports, and airports.

Advantages to the owner

Concrete pavers are available in hundreds of different product lines. These simulate the appearance of other materials such as bricks and fitted stones or have new, novel, and even whimsical appearances. They offer a broad range of premium design options.

Pavers are durable and low-maintenance. They can adjust to the loads above and slight ground movement without cracking. They resist fuel and oil as well as freeze-thaw damage. When repair is necessary, it is relatively quick and simple because the problem pieces can simply be removed and replaced.

Advantages for the contractor

The installation of concrete pavers provides an added high-end, high-margin business for landscape contractors. There are only few simple tools involved in small installations. Changes or adjustments are relatively easy at almost any point of the assembly process. Although some knowledge is necessary for quality work, it is easily learned, and training and support information are readily available.

Components

Most installations involve the pavers themselves, then up to four layers beneath them. The layers below are called the *soil subgrade, subbase, base,* and *bedding sand.* An *edge restraint* is also usually installed. Also used are *joint sand* and, in some projects, a geotextile.

Concrete pavers are flat, solid concrete units in various shapes. Two standard thicknesses are available. For pedestrian and residential driveway applications, pavers 2⅜ inches thick are recommended. Thicker, 3⅛-inch units are used for regular car and truck traffic.

The shapes of the pavers are generally designed to fit one or more of four different paving patterns: *running bond, 90-degree or 45-degree herringbone, cross joint bond,* and special patterns, some of which are suited primarily for automated installation on larger projects with specialized equipment. In some patterns there are also special edge pieces (Figure 28-2).

The subgrade is usually compacted soil. A subbase typically consists of small stone. The base is a dense aggregate like coarse sand. Bedding sand is finer. Edge restraints are typically plastic or metal strips. In some projects long, narrow stones may be used.

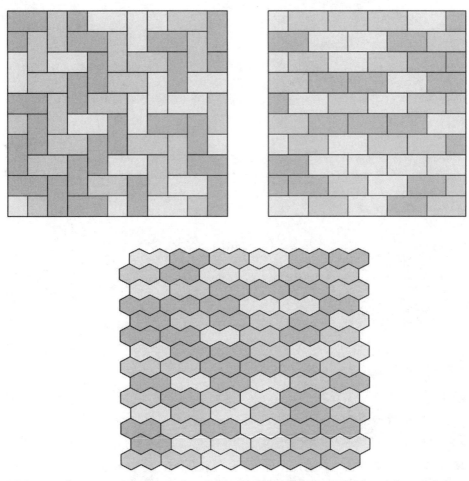

28-2 *Popular paving patterns: herringbone (top left), running bond (top right), and a special shape placed in running bond (bottom).*

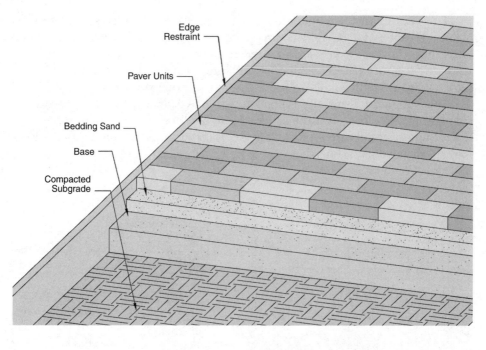

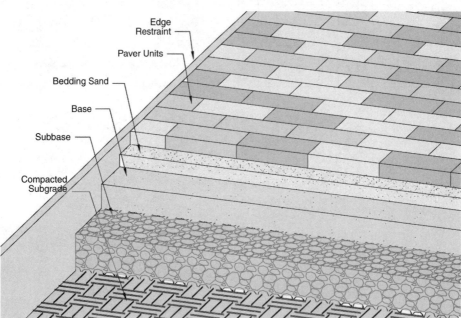

28-3 *Cutaway view of a paver installation in a typical medium-duty application (top) and in a heavy-duty application in unstable soil (bottom).*

Pavement assembly

A quality paver assembly includes layers of other materials below grade. The number and depth of the layers depend on the application and the quality of the existing soil.

Nearly all installations use the natural soil compacted for the subgrade. Geotextile typically goes over this in the case of silt or clay soils. Above this is a base layer of about 4-6 inches of crushed stone. This is compacted and leveled. The base extends several inches beyond the planned edges of the paver installation. The edge restraints are on top of the base. Edge restraints are along all sides, except where there is already a side support such as a wall or curb. A layer of about an inch of bedding sand lies on top of the base, inside the restraints.

The pavers are on top, arranged according to their paving pattern. On the opposite face of the edge restraints is soil. Between the pavers may be joint sand. The pavement normally has a slope of at least 1.5 percent ($3/16$-inch drop per foot of surface) for drainage. The slope is recommended not to be over 12 percent to avoid instability (Figure 28-3).

For heavy-traffic projects or where the soil is unstable, it is common to add a special subbase. The subbase may have geotextile beneath it and may be several inches deep. The rest of the assembly above is the same.

Permeable pavements involve different base materials and thicknesses that vary with the infiltration capacity of the soil. These pavements typically have little or no slope and use special units that have small openings to allow runoff to enter and drain into the base and soil.

Cost

Landscape contractors charge $7.00 to $10.00 per square foot for a typical paver installation on a small building lot. This assumes installation for medium-duty purposes on stable soil. The quote includes the subcontractor's markup. The costs of construction are approximately as in Table 28-1. Figures in the table do not include a subcontractor markup.

Table 28-1. Approximate Costs of a Paver Installation.

Item		Cost per Square Foot*
Labor		$2.50
Materials		$4.05
Pavers	$3.50	
Base and bedding sand	$0.20	
Edge restraints	$0.35	
Equipment and fuel		$1.00
Total		$7.55

*All costs in U.S. dollars per gross square foot of pavement area.

A very simple gray paver could reduce costs by a dollar or two per square foot. The most elaborate pavers could raise it as much as four dollars. If a special subgrade and a subbase are required, cost could increase about two dollars per square foot. Most of this would be for the earth moving to remove the existing soil and replace it with more stable material. Complex projects are generally more costly. Complexity comes from many angles or curved borders, changes in elevation requiring steps, or mixing different paving patterns. Much of the higher cost of complex layouts comes from the need to cut pavers, which is a relatively expensive operation. Labor rates also vary regionally, with costs generally less in the southern United States.

Code status

Paving is generally not closely regulated in small projects. Inspection by the local building department is also rare.

Pavers in the United States are commonly expected or required to meet the standard ASTM C936, "Standard Specification for Solid Interlocking Concrete Paving Units." This requires the pavers to have average compressive strength of 8,000 psi, water absorption of less than 5 percent, and good resistance to freeze-thaw cycles.

In Canada the relevant standard is "Precast Concrete Pavers" from the Canadian Standards Association (CSA). It has similar requirements.

Installation

Proper installation begins with determining the quality of the soil and the likely weight of the traffic. This determines whether a special subgrade or subbase will be necessary and how deep they must be. It is also important to make sure that there are no utility lines under the installation that might be affected by digging.

Work begins by staking the perimeter and establishing elevations. This includes any slope for drainage. The ground is excavated to a depth equal to the total depth of all layers. The excavation should extend 12 inches past the edges of the paving on all sides.

Each layer of material is deposited into the excavation and compacted. The compaction is typically done with a vibrator plate (Figure 28-4)

After the subgrade and subbase (if any) are in position, the base is deposited and compacted. After that, the edge restraint is installed. The crew then fills the excavated area outside the paving area to back up the edge restraint. The fill is normally topsoil.

If bedding sand is necessary, it goes on top of the base.

The crew sets the pavers according to their paving pattern. Normally, work starts at the center and proceeds toward the edges. It also begins at the low-

28-4 *Leveling the base sand.* Interlocking Concrete Pavement Institute

est point and proceeds uphill. The workers maintain a gap of $\frac{1}{16}$ of an inch to $\frac{3}{16}$ of an inch between the units. For a quality installation the crew maintains precise spacing and layout with string lines.

Spaces at the edge are filled with edge units or cut pavers. Cutting can be done with a special tool called a *paver splitter* or a masonry saw. Cutting off thin pieces is avoided, as they can crack.

When the pavers are down, the crew sweeps joint sand into the joints. They then compact the pavers into the bed below with a plate vibrator. After this compaction they sweep more joint sand over the surface and compact again until all joints are full and the pavers are securely in position. This generally requires two or three passes. Then the crew sweeps off the excess sand (Figure 28-5).

If work takes over a day, it should stop with enough time to compact the set paving. That compaction should stop 3 feet from the unrestrained edge of the pavers. If there is a chance of rain, uncompacted pavers are covered with plastic sheet.

Pervious paver installation is the same, except that the pavers are shaped in a way that leaves larger regular gaps. The sand used may also be coarser to allow easier drainage.

28-5 *Worker sweeping sand into the joints between pavers.* Interlocking Concrete Pavement Institute

Connections

Increasingly, pavers are installed right up to segmental retaining walls. In this case, the areas for the paving and the wall are both excavated. The wall is constructed first, and the base for the pavers is prepared right up to it. The pavers are then installed as usual up to the wall. The wall acts as the edge restraint. The pavers may be set outside the wall right up to the base, or inside the wall right up to the head.

Architects

Most landscape architects are now familiar with concrete pavers and capable of designing with them. However, if an architect has done few paver projects, it is advisable to set up early meetings with the installer. Many angles, curves, odd dimensions, and elevation changes lead to cutting and labor. They can affect cost and schedule in ways that the architect may not be able to predict accurately.

Finding a qualified landscape architect usually involves reviewing some listed in local directories or the American Association of Landscape Architects (ASLA) web site at www.asla.org.

Engineers

Engineers are typically involved in heavy applications. They are rarely involved in a small building project.

Retaining a civil engineer may be recommended or required to determine the type and depth of subbase and base layers to use, particularly in a case of poor soil. *ICPI Tech Spec 4 Structural Design of Interlocking Concrete Pavements* from the Interlocking Concrete Pavement Institute (www.icpi.org) provides information on this task for the engineer.

Training

Although many landscape contractors learn installation from the manufacturers' manuals, it is generally recommended that they attend a formal training class. This can lead to a shorter learning curve and better quality work with lower costs.

ICPI offers a two-day installer certification course through its members across the U.S. and Canada. The course locations are listed on the ICPI web site, www.icpi.org. Check with local ICPI member paver manufacturers for most current information on certification classes.

Maintenance and repair

Very few owners perform any routine maintenance on their paving. Wear is simply too slow to justify it.

Very occasionally a unit may crack, or some may wear over time so that their appearance is no longer acceptable. These can be simply replaced by removing the old ones, taking out any excess sand, setting in new pavers, and filling the joints and compacting. If depressions develop over time, the low units may be removed, extra sand put down, and the units reinstalled.

Sealers can be applied to the surface of pavers to resist oils and dirt and make cleaning easier. Sealers will need to be reapplied. The application rate depends on the climate and the amount of wear from traffic. Detailed information is available from ICPI (www.icpi.org) and the sealer manufacturers.

Sustainability

Depending on their color, pavers may reflect sunlight and prevent the buildup of heat on the property. Pervious pavers allow water to drain into the soil below and reduce the runoff and the erosion and other problems arising from it. In fact, pervious pavers are often used as an integral part of runoff control strategies on heavily paved sites. Each of these benefits is recognized by Leadership in Energy and Environmental Design (LEED) and may contribute to the achievement of a point on the LEED rating scale.

Pavers are almost always locally supplied. They may also have some recycled content, although this is rare and must be checked with the manufacturer. Either of these properties may also help qualify for a LEED point. Because the pavers are often a small percentage of the whole project, these attributes may not make as great a contribution to achieving a point.

Aesthetics

Concrete pavers are available in countless patterns that interlock. Most consist of geometric shapes.

The available colors include almost anything. Most often muted tones are chosen to blend better with the soil that inevitably gets on all pavement. Partial blends of two colors produce veining effects or a mottled appearance. These are also popular.

The typical finish is simply flat with a fine sand texture. Some units are now available with a slightly irregular, random surface that resembles rough natural stones or rounded edges that resemble aged paving stones.

Key considerations

Key to a sharp, long-lasting paver installation is proper preparation of the base layers below it.

It is also important not to install in conditions where water could cause shifts in the soil or the base layers. This includes installing during heavy rain or snowfall, over frozen base materials, or over frozen or saturated sand.

Availability

There are now many paver producers across North America. To find a local one, check in the ICPI directory of members (on www.icpi.org) or local listings.

The base materials are standard and available all over North America. They are the same as those used under asphalt paving. Geotextile is a less common item. It is often available from the paver supplier. If not, that supplier should be able to direct buyers to a source. It is not advisable to use landscape fabric meant for preventing weeds. This material is not of the required strength.

Crew availability

Most landscape contractors install pavers or can direct buyers to someone else who does. Experienced contractors are located in nearly every significant region of the United States and Canada. Simply contact landscape contractors in public listings and ask for credentials, or find contractors who are members of ICPI, by searching the directory at www.icpi.org.

Support

Experienced contractors rarely need help. However, they can get basic assistance and usually detailed literature from the paver supplier or the ICPI (www.icpi.org)

Current projects

To find a current project to view, ask local landscape contractors what they have coming up.

29

Flatwork

Concrete flatwork consists of a layer of concrete cast on the ground to a thickness of about four inches. It creates a solid, durable surface for driving, walking, and enjoying the outdoors. Because the concrete is placed in a liquid state, it can be formed into almost any perimeter shape desired, to conform to the site or create an artistic outline (Figure 29-1).

Concrete flatwork has a long track record of success. It is widely available and commonly understood. Estimates are that over one hundred million square feet of flatwork are created in North America each year on the lots of small buildings. Data indicate that over half of all driveways, a little under half of all patios, and 80–90 percent of all walkways are built with concrete flatwork.

With the development of technology, flatwork has new capabilities as well. The advance of decorative concrete has opened the potential for high-design surfaces outdoors similar to the beautiful floors now being created indoors. Ingenious new concrete mixes allow exterior flatwork to be pervious. Pervious flatwork allows water to drain through it and pass into the soil instead of flowing over the ground as runoff that can create erosion and other problems. These specific topics are covered elsewhere in the book. Decorative concrete for floors and flatwork is covered in Chapter 34. Pervious concrete is covered in Chapter 31. Both are variations on the basic concrete flatwork described in this chapter.

History and market

Like so many other aspects of concrete construction, concrete flatwork owes much of its development to the Romans. They cast their version of concrete on the ground to create stable, clean, durable common areas and roadways.

With the development of modern cement, North Americans began constructing roads with concrete placed on the ground in the late 1800s. This replaced roads that were usually nothing more than compacted dirt or stones set on the ground. This use has continued and grown steadily to this day.

29-1 *Placing concrete for a driveway.* Portland Cement Association

Over time, the material and the construction process have become more economical. Flatwork became common in all parts of the construction market as an inexpensive, durable alternative to asphalt paving. Today, most houses and small buildings in all segments of the market have concrete flatwork outdoors. They are used to create walks, drives, and patios.

Advantages to the owner

Concrete is a highly rigid material that resists deforming or compressing in any weather. It has a durable surface that resists fuel, oil, and freeze-thaw damage. It is also usually light in color. This reduces heat buildup on the lot during hot, sunny days.

Nowadays concrete flatwork can also be given new, important properties. One is a decorative concrete surface. This is discussed further in Chapter 34. Another is permeability to create pervious pavement that allows rainwater to drain through, directly into the soil. This is discussed further in Chapter 31.

Advantages to the contractor

Concrete flatwork is widely available and well understood. It is easy for the contractor to work with. On most projects there is other concrete work on the site as well. The flatwork can often be created in coordination with this other work, producing efficiencies.

Components and pavement assembly

Concrete flatwork is made with conventional concrete and reinforcement. The reinforcement is typically welded wire mesh or a reinforcing fiber that is mixed into the concrete.

The slab rests on a subgrade. This may be simply the soil compacted down. If the soil is unstable material, a more solid material such as crushed stone may be laid down in a shallow layer. The ground is cut to a depth that will bring the top of the flatwork to the desired level after the subgrade material (if any) and concrete are added.

The flatwork is normally $3\frac{1}{2}$ inches to $4\frac{1}{2}$ inches thick. The greater thickness is typically for heavier use applications like driveways.

The perimeter can have virtually any shape. The concrete surface is to be sloped $\frac{1}{8}$ inch in every foot away from the building. This prevents water pooling next to the building. Welded wire mesh is embedded close to mid-depth in the slab. The edges of the mesh should end an inch away from the edge of the slab (Figure 29-2).

Many installations include control joints every few feet. These are linear depressions in the surface, about one inch deep.

Cost

Concrete contractors charge about $4.00 per square foot for an installation of plain flatwork. This includes the subcontractor's markup. The costs of construction are approximately as in Table 29-1. Figures in the table do not include a subcontractor markup.

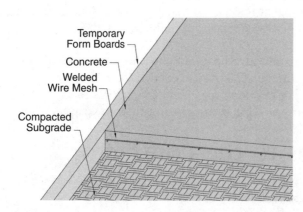

Temporary
Form Boards

Concrete

Welded
Wire Mesh

Compacted
Subgrade

29-2 *Cutaway view of a typical flatwork installation.*

Table 29-1. Approximate Cost of Flatwork Construction.

Item		Cost per Square Foot*
Labor		$2.00
Materials		$1.25
Concrete	$1.00	
Reinforcement	$0.25	
Total		$3.25

*All costs in U.S. dollars per gross square foot paving area.

Costs can be higher if the shape of the project is complex. Odd shapes typically require more labor and possibly the purchase of some form materials that cannot be reused. Cost is also higher if special material must be brought in to provide an adequate sub-grade. This can require more digging to make room for the material, as well as the cost of the material itself and the labor to deposit it.

Code status

Paving is generally not closely regulated in small projects. Inspection by the local building department is rare.

Installation

Installation begins with preparing the subgrade. If necessary, the ground is cut to depth so that the finished flatwork surface will be at the desired height. If the soil is stable and it was not disturbed, then further preparation may not be necessary. If it has been disturbed, the crew may compact it with a vibrator plate or other device. If it is unstable, the crew places a layer of stone or other material and compacts it into the soil (Figure 29-3).

The forms are staked around the intended perimeter of the installation. The forms are typically 2× dimensional lumber set on edge and staked to the ground. Metal and plastic forms are also available. Some of these are designed to flex so that they can create curves. Curves may also be formed with thinner sheet material backed by frequent stakes.

With the forms in place, welded wire mesh can be set. This is frequently installed on chairs or other supports to hold it at mid-slab depth. Alternatively, the project may use fiber reinforcement in the mix.

The crew most often places the concrete simply by directing the chute from the concrete truck inside the forms. They spread the piles of concrete evenly throughout the formwork with rakes or similar tools. A vibrator may be used to consolidate the concrete. The crew then finishes it by floating. If a slightly rough finish is desired for traction, they typically draw a heavy broom across the surface (Figure 29-4).

29-3 *Compacting the subgrade with a vibrator plate.* Portland Cement Association

29-4 *Floating flatwork.* Portland Cement Association

Control joints may be created while the concrete is still soft by drawing a hand tool across the surface to the proper depth. They may also be created after the concrete has cured by making shallow cuts in the slab with a concrete saw.

Architects

Architects and landscape architects are thoroughly familiar with concrete flatwork. They generally need no particular assistance to design or specify basic flatwork.

Engineers

Engineers are rarely involved in the design or specification of flatwork on small buildings. In the case of very heavy applications, they may be retained to specify critical structural items like concrete strength, slab thickness, and reinforcement.

Training

Formal training is not normally required for small-project installation, and many workers learn their trade assisting on established crews. But formal courses are available. Training of the crew supervisor can increase productivity and work quality.

Courses in flatwork are offered at major trade shows, such as World of Concrete (www.worldofconcrete.com) and others. The American Concrete Institute (ACI) offers the Concrete Flatwork Finisher and Technician (CFFT) Certification for contractors who demonstrate a high level of knowledge and skill. Two certification levels are available. The CFFT Certification requires passing a written examination. The higher CFFT Finisher Certification requires passing a written examination, a performance examination, and a least one year of flatwork experience. These examinations are also administered at major trade shows. See www.aci.org for details.

Maintenance and repair

Concrete flatwork can last decades without maintenance.

Fine cracks can form in some cases from shrinkage during concrete curing. The solution for this is prevention. Shrinkage cracks are minimized with a mix that has a low water-cement ratio, proper reinforcement, and control joints.

Later cracking is sometimes caused over time by the freezing of water that penetrates the pores of the concrete. This is also called *spalling*. If it is severe, it can be repaired by chipping out the damaged area and filling it with a patching compound. Patching compounds are concrete mixed specifically for filling holes. They are available ready-mixed, or can be made by combining certain standard ingredients.

Sustainability

The concrete used in flatwork almost always comes from relatively near the site. Depending on the amount of material involved, this may help qualify the project for the Leadership in Energy and Environmental Design (LEED) point for local production. Flatwork concrete can sometimes include recycled material. If so, this may help a project qualify for a LEED credit for the use of recycled materials. It would be necessary to contact the manufacturer to determine what recycled content the concrete may contain. Since concrete is light in color, it reflects most sunlight and keeps the site cooler. This can be instrumental in helping a project qualify for the "heat island effect" LEED credit.

Aesthetics

Basic flatwork installation produces a standard gray surface. It can be varied in shape through the use of framework. It can be given dramatic and varied surface textures and colors through the use of decorative concrete techniques. These are covered in Part 6 of this book.

Key considerations

The key to good concrete flatwork is in proper materials selection and installation technique. Trained and experienced contractors are the best method of insuring quality work in these factors.

Availability

The components of concrete flatwork are available from concrete ready-mix suppliers and general building supply across North America. There are tens of thousands of concrete contractors that create flatwork. They can be found through local business and construction listings.

Support

Experienced contractors rarely need help. They may get advice from their concrete suppliers and from the suppliers of chemical admixtures for concrete.

Current projects

Flatwork projects are constantly underway across North America. They can be located by contacting local contractors to ask about their upcoming projects.

30

Segmental retaining walls

Segmental retaining walls (SRWs) are made of special concrete blocks stacked directly on one another to hold back the earth. Retaining walls in general have become increasingly important on the lots of small buildings. Growth in construction has driven us to build on lots with sloped land. The retaining wall helps make use of the land by dividing it into two flat areas instead of one sloped one.

Segmental retaining walls add new dimensions to the traditional retaining wall. They are designed for rapid installation with a minimum of materials. The blocks are simply set on top of one another. This makes SRWs much more economical and practical than older methods of retaining wall construction. The blocks can also be colored and textured in a variety of ways to create a much more decorative wall than the average buyer could afford previously. This has led many to create walls in situations where they were not strictly necessary, as a decorative element on the lot (Figure 30-1).

Invented only 20 years ago, SRWs have taken off. Although exact figures are not available, they are thought to sell well over 100 million square feet per year in the United States and Canada. On small building lots SRWs alone now probably account for more retaining wall construction per year than was built with *all* methods twenty years ago.

History

Piling up natural stones to create a small wall with different elevations on each side appears to be so old that it is difficult to say when it started. Retaining walls made of stones became popular in Europe during the 1700s. They caught on in the United States about a hundred years later.

By the 1970s a variety of other methods were also used to create retaining walls. The lower walls typical of small lots were often built with 8 × 8 or

30-1 *Segmental retaining wall.* National Concrete Masonry Association

6×6 timbers. They were set sideways on one another and often connected with large bolts. They were typically arranged into a wall with a backslope, which was an easier elevation for the light timbers to hold. Some retaining walls were constructed with standard concrete block by a mason and some were reinforced concrete built with removable forms. A great many were still built with natural stones. These might be found at the site during excavation and set to one side. When it was time to build the retaining wall, the workers with the earth moving equipment piled them on one another and placed the earth behind them. Some skill was required to get the irregular stones to perch securely and with the desired slope.

In the 1970s a company in Virginia began offering precast panels designed to be set in place to hold the earth back. In the 1980s, inventors in Minnesota came up with the idea to create smaller units on a concrete block machine and designed to be stacked on one another to make a retaining wall. Unlike natural stones, they were all one size and fit together precisely. They had knobs or pins to hold one on the other without sliding off. Because of smaller pieces, interchangeable parts, and precise fit, crews could create a stable wall rapidly with limited use of heavy equipment.

By the 1990s there were several SRW companies with somewhat different designs. They licensed manufacturing rights to concrete block producers across North America.

Two important developments arose during that period. One was the use of geotextile to help stabilize the soil. This is actually a variation on an old practice. The Egyptians used papyrus for the same purpose and the Chinese stabilized the soil with strips of bamboo when constructing the Great Wall. Laying layers of geotextile in the ground behind an SRW stabilized it so that the blocks could be built higher and in places with unstable soil.

The second development was growth in the use of decorative colors and finishes. By taking advantage of the new decorative concrete technologies, retaining wall units have been given appearances ranging from aged, weathered stones to modern, high-tech materials.

Market

The sale of SRWs is now spread across all regions and all markets. Initially, sales were concentrated in the Upper Midwest, around Minnesota where the first SRW companies were located. However, they are now marketed across the United States and Canada. They appear wherever there is construction.

Segmental retaining walls naturally sell more heavily in regions with uneven terrain for the purpose of making flat stretches of land. They sell to all segments of the market for this purpose. They are nearly a necessity on uneven lots for homes and commercial, institutional, and industrial buildings of all types.

In addition, nowadays SRWs often sell for decorative and enjoyment purposes. Upscale home projects and commercial properties often use SRWs to create interesting patterning and secluded areas on their lots. These sales can be in any region and more often involve premium-priced, highly decorative units.

Advantages to the owner

Segmental retaining walls serve their purpose attractively and without maintenance. The walls have very good flexibility. The material can form walls from eight inches to many feet tall. The walls can be installed to virtually any straight or curved plan. A wide range of colors, finishes, and types of appearances is available. Segmental retaining walls resist deterioration from the elements. Properly installed, the walls can be expected to last at least a century without maintenance.

Advantages for the contractor

On a small project, segmental retaining walls are generally easier to install than any other retaining wall system. In small projects, they can be installed with no heavy equipment. The work is readily learned and does not require a great deal of skilled labor.

Components

The major component of a segmental retaining wall is the SRW masonry unit. The SRW units for small projects range in weight from about 20 pounds to 100 pounds each. They may be solid or hollow. They also have some means of interconnecting with other units above and below them. In some cases they have holes and lugs or ridges positioned so that the lug/ridge in one unit fits into the hole on the one above or below. Other designs have only holes in the units, but also have plastic *pins* that fit into the holes to connect a unit below with another one above (Figure 30-2).

Most SRW systems also have a solid, flat unit called a *cap*. Typical dimensions for a cap unit are 2 in. × 8 in. × 16 in. Some SRW installations also use geotextile fabric. Some use *drainage stone*, which is typically ordinary crushed stone of about ½ inch to 1¾ inches in size.

Wall assembly

The typical retaining wall on a small lot is 2 feet to 4 feet high. Below the bottom of the wall is a base of about 6 inches of sand or crushed stone. The seg-

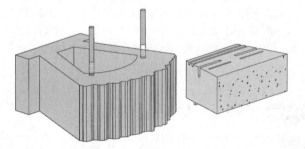

30-2 *An SRW system connecting its units with pins (left), and one with lugs (right).*

mental retaining wall units are stacked on top. They are almost always arranged in a running bond pattern, which is described in Chapter 6. Hollow units are filled with drainage stone. Solid units typically have a layer of drainage stone behind them about 12 inches wide.

The units are connected to those above and below by the fit of lugs or pins into the holes. For very short walls, the units may line up exactly vertically. For some short walls and all taller walls, they are stacked so that each higher course is set back about $\frac{3}{4}$ inches to create a wall that inclines back toward the retained earth. Most units are designed with holes positioned so that they can be stacked with this slight setback (Figure 30-3).

For taller walls, there is typically one layer of geotextile installed every four feet up the wall. Each layer lies horizontally in the retained earth. The front edge of the geotextile extends into the joint between two courses of the SRW units. The geotextile is held there by the plastic pins, which are run through the fabric, or by some other means. The geotextile extends back into the earth several feet.

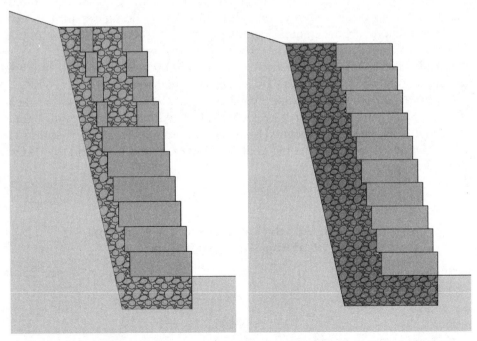

30-3 *Cutaway view of a segmental retaining wall built with hollow units (left) and one built with solid units (right).*

Cost

Many retaining wall projects have more cost involved in earth moving than in the labor and materials to create the wall itself. Landscape contractors typically charge $10.00 to $15.00 per square foot for the wall, without excavation. The quote includes the subcontractor's markup. The costs of preparing and finishing the earth may double or triple this.

The costs of wall assembly for a low retaining wall using basic units are approximately as in Table 30-1. The figures in the table do not include a subcontractor markup or excavation.

A very plain SRW unit might be less expensive. Highly decorative units would be more. Intricate walls with many angles or turns will also be more expensive. If a geotextile is needed, this may add about $2.00 to $4.00 per square foot for the fabric and the labor to place it. The excavation and backfill costs vary widely. They depend on how much earth must be moved and whether special material is necessary behind the wall.

Code status

Low retaining walls are generally unregulated in the building codes. Local building departments do not often inspect the walls. The building department may place restrictions on the general contour of the lot to control such things as runoff. The use of retaining walls may be part of this.

Walls taller than 4 feet usually require a building permit, and the building department may require analysis by a licensed engineer.

A formal standard for retaining walls is ASTM C 1372, "Standard Specification for Segmental Retaining Wall Units." Most manufacturers produce their units to this standard. It is designed to insure performance, durability, and structural integrity. It regulates the materials used to make the units and the design of the system.

Table 30-1. Approximate Costs of a Segmental Retaining Wall Installation.

Item		Cost per Square Foot*
Labor		$2.00
Materials		$8.50
SRW units	$8.00	
Crushed stone and sand	$0.50	
Total		$10.50

*All costs in U.S. dollars per gross square foot wall area.

Installation

The landscape crew excavates to a little below the bottom of the planned wall. Many systems call for the bottom course of units to be below grade. They excavate back into the hill far enough to create the desired ground elevation. When using a solid block, the cut back from the planned rear of the wall must be at least 12 inches to allow room for the drainage stone.

Below the wall the crew installs the bed of sand or stone and compacts it. This may be done with a vibrator plate.

If the wall has a corner, installation of the units starts there. The units on the first course must be level across their tops and from one to the next. They may be set to a string line to insure that the wall is straight (Figure 30-4).

After the first course is set, the crew pushes the plastic pins (if any) into holes in the top of the units. The next course is set on top, fitting the top units over the pins or lugs on the units below.

Drainage stone goes into the cavities of hollow units. It goes behind the wall when using solid units. Some systems call for placement of stone after every course, and some after every four feet of wall height. When placing the stone, the crew also places backfill behind the stone to fill any further excavation.

The backfill behind the wall is typically compacted every few feet. For a tall wall, geotextile is placed when the wall reaches four feet. The crew sets

30-4 *Completed wall with caps attached.* National Concrete Masonry Association

it over the pins or lugs protruding from the top of the units and lays the rear section on the stone and backfill behind the wall.

The process of placing courses of units, backfilling, and layering geotextile on top continues until the wall is at the desired height. The crew affixes cap blocks to the top course of units with a concrete adhesive. They may overhang the front edge of the units below, be flush with them, or be set back slightly, as desired.

Connections

As noted in Chapter 28, pavers are sometimes installed right up to segmental retaining walls. There is, however, no physical connection between them. The wall acts as the edge restraint for the paving. The pavers may be set outside the wall right up to the base, or inside the wall right up to the head.

Architects

Because SRWs are now the most common method of creating a retaining wall, nearly all landscape architects are familiar with them. There may be some value to putting the architect on a particular project in contact with the contractor, so both are in agreement as to what types of systems are required and available. However, rarely does the architect need any training or education on the product. Landscape architects can be found in local business and construction directories.

Engineers

Engineers are required to help design tall SRWs. These are rarely found in small building projects. However, building departments often require engineering for walls taller than four feet. The engineering is usually fairly simple and well documented in the technical literature from the SRW company. Not all engineers do SRW design, but most local areas have several. They can often be found by asking a landscape architect, landscape contractor, or SRW supplier.

Training

Formal training is not generally required to become an SRW installer, and many learn from the product manuals. However, training classes are available and many have commented that they bring the installer up to speed faster, make the work more efficient, and help keep quality up. The courses are offered by the sellers of the products. This includes the companies that designed the SRW systems and the block producers who manufacture and sell them. Information is available from these companies, usually on their web sites. Many SRW companies and manufacturers are members of the National Concrete Masonry Association (NCMA). They can be found by searching the Membership Directory of the Association, www.ncma.org.

Maintenance and repair

Well-installed SRWs should last decades or even centuries without maintenance or repair. Some sellers have offered lifetime guarantees. In areas with high water flow the surrounding soil may eventually erode and need to be replaced. The units will age and wear gradually over time, like any masonry or stones. Water flows may carry minerals that affect the appearance of the units after a period of years. If this is considered undesirable, the units may be replaced.

If the drainage layers of the wall were not properly installed, the fluid pressure behind it may eventually push the wall out. However, this is rare. It also tends to be less common with SRWs than with some other types of retaining walls because SRWs are heavy, they have high friction, and they are designed to allow water to drain.

Sustainability

Segmental retaining walls have some properties that may be related to the environmental impact of a project. However, they do not usually figure heavily into measures of sustainability

Segmental retaining wall units are almost always locally supplied. This may help a project qualify for the Leadership in Energy and Environmental Design (LEED) point for use of locally produced content. They may have some recycled content, although this is rare and must be checked with the manufacturer

Aesthetics

Segmental retaining walls are available in a wide range of colors, finishes, and styles. The most popular colors are earth tones, to provide a natural stone look. However, any others available in concrete masonry can be produced and sometimes are.

Split face is the most popular finish. It also provides a stone-like appearance with a rough, fractured texture. Increasingly popular is the *tumbled* or distressed unit. These have intentionally worn, rounded edges that make them resemble old stones that have been smoothed by wear and the elements.

The units are also available in many sizes and basic shapes that give them sharply different appearances.

The types of units available differ from one supplier to another. Most areas are now served by several suppliers, so it may be worthwhile to compare.

Key considerations

The precision of an SRW installation depends on creating a firm, level base for the wall. The first course should be set precisely, since it dictates how the upper courses will align.

One of the advantages of the segmental retaining wall over some other systems is that it is designed to drain effectively. The pressure of water behind the wall is the largest force acting on most retaining walls. Over time many other types of walls begin to lean outward and must be replaced. So that the SRW can avoid this fate, it is important that the drainage stone be properly sized and installed.

Availability

Today, nearly all localities across North America have multiple sellers of SRW units. Perhaps half of all concrete block manufacturers now sell them. These can be found in local business directories or by searching the directory of members of the National Concrete Masonry Association (www.ncma.org). Segmental retaining walls are also available from most home centers, which are a practical source for smaller orders.

Crew availability

Segmental retaining walls are now such a standard element of landscaping that most landscape contractors install them. The contractors themselves can be found in local business and construction directories.

Support

Most contractors require little support on small walls. When they do, it is usually available from their local SRW supplier. For certain detailed questions it may be necessary to contact the company that designed the system.

Current projects

To find a project to view, ask local landscape contractors what they have coming up.

31

Developments in landscape products

Turf pavers

An interesting new type of paving unit is the *turf paver*. It creates a more natural looking landscape, allows free drainage of water, and helps keep the surrounding area cool in hot weather. The turf paver is a large, flat concrete masonry unit with a series of holes in it. When it is installed, plants grow through the holes to create a sort of lawn that is also suitable for traffic (Figure 31-1).

Turf pavers are installed directly in the topsoil. Plants are allowed, or even encouraged to grow through the holes. Various types of plants can grow there. Typically, owners plant grass. Normal traffic keeps the plants short. If necessary, they can be cut with most lawn mowers.

Turf pavers are suitable for creating walks and drives. They can support cars, and they provide a stable surface that is largely free of mud. Yet, from an angle the paved area looks like a lawn.

The paving has strong environmental advantages. It is highly pervious. Water flows freely through the holes and into the earth below. This reduces runoff and its harmful consequences. During hot weather, the moisture in the earth can evaporate, cooling the area. The lawn also remains a light color that gains less heat from the sun, unlike some solid paving.

Turf pavers are available from many of the suppliers of conventional concrete pavers, and can be installed by many landscape contractors.

Pervious concrete

It is now possible to buy ready-mix concrete that water can pass through. This makes it possible to place liquid concrete on-site into formwork to create a walk or drive of any shape, and still have paving that drains water through to the soil without runoff (Figure 31-2).

31-1 *A turf paver installation.* E.P. Henry Company

31-2 *Water draining through pervious concrete.* Portland Cement Association

The new *pervious concrete* is made with special mixes that include no fine aggregate such as sand. The cement coats the stones in the mix and connects them to each other, but with considerable gaps as well. The concrete cures this way, and water can pass through the gaps. Various methods can be used to clean out the gaps if they become filled with dirt or the like.

Pervious paving is relatively new. Only some ready-mix suppliers offer the concrete. Where it is available, it provides the advantages of low runoff with the installation economy and flexibility of site-cast concrete. Contact local ready-mix suppliers for information.

Combined paver-SRW installations

A growing number of landscaping projects are using concrete pavers and segmental retaining walls in combination to create features that were formerly built of less decorative and less durable materials.

Usually the SRWs are arranged to create an elevation and the pavers form a platform at the top of the elevation. For example, some owners who would otherwise build a wooden deck outside their homes have instead installed a "raised patio." Like a conventional deck, its surface lies about a foot above grade. However, it has sides consisting of a low segmental retaining wall. Its top surface consists of pavers. Naturally, fill has gone inside the perimeter of the retaining wall, and the pavers rest on that. The result is a highly attractive deck with the long durability and low maintenance of concrete (Figure 31-3).

31-3 *Segmental retaining walls and pavers combined to make an outdoor "deck."* Portland Cement Association

The front steps to a home have also been created by this method. There are also arrangements where the retaining walls create a lower area that is covered by pavers, such as an activity pit. There are of course also many installations that include SRWs and pavers that are not joined for any special function, but are coordinated with one another for aesthetic purposes.

This appears to be part of a larger trend of experimentation with these landscape products. Pavers and SRWs are valuable new tools for landscape designers. The designers are using those tools to develop a broad range of new applications and benefits for customers. It appears that the only limit is the imagination.

PART VI

DECORATIVE CONCRETE

32

Background on decorative concrete

Decorative concrete is not so much a product as a new wave of interior design and a set of techniques. These techniques are surface treatments and forming methods that can make any of a wide variety of concrete products strikingly beautiful and unique. Decorative concrete has opened up a whole new interest in concrete as an aesthetic material, prized for the striking and novel forms and appearances it can have.

The most common uses of decorative concrete are for countertops, floors and exterior flatwork. However, a major result of the decorative concrete movement has been that the practitioners constantly experiment with making more and different parts of the interior out of the material.

Concrete countertops were almost nonexistent before the current wave of decorative concrete began in the 1980s. They are valued and purchased largely because they are treated with decorative concrete techniques.

Floors and exterior flatwork were already widely used before the new techniques came along. However, they gained a whole set of new finish options when these techniques were applied to them. They are still often left without decoration. The construction of basic floors and flatwork have already been covered in this book in Part 3 and Chapter 29, respectively. Most floors are covered with carpet, tile, and the like. Most flatwork is still left smooth and gray. However, a rapidly growing number of owners are choosing to adorn surfaces with decorative concrete treatments.

One of the attractions of decorative concrete is that it makes a home or building unique. Virtually all of the installations are custom designed and custom made for the specific project. Decorative concrete is a grass roots industry of artists, with hundreds of independent shops and individuals producing individual works of art. Experimentation continues, with new finishes, methods, and applications appearing constantly (Figure 32-1).

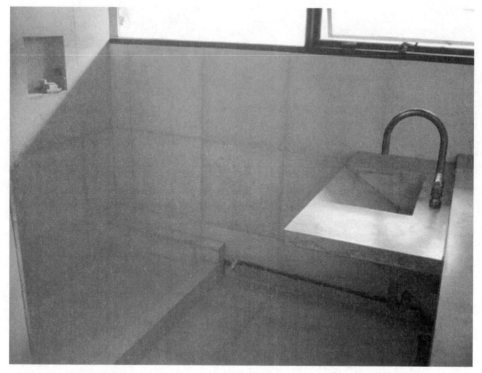

32-1 *Bathroom with decorative concrete counter, sink, floor, walls, and fixtures.*
Stone Soup Concrete

Decorative concrete has some significant differences from the other products and systems in this book. Its function is aesthetic, and not a matter of safety, structure, or protection for other parts of the building. It therefore does not usually need to be engineered, regulated or inspected.

In many parts of a decorative concrete project, how well the work has been done is a matter of opinion. The primary purpose is beauty, and beauty is in the eye of the beholder. It is possible to make recommendations on materials and workmanship that influence other important properties such as durability and cost. However, for the key aesthetic decisions of a project it is not possible to identify a "right" or "wrong." The supervisor and customer instead are giving over significant control of the project to an artist. In return, they hope to receive a beautiful work of art that comes from creativity.

History

The decoration of floors dates back to ancient civilizations that applied paint or dyes or indentations to add appeal. After the arrival of modern cement in the 1800s, there also appeared various simple means of adding texture to flat horizontal surfaces like sidewalks. These include such old, established methods as the broom finish or salt finish.

During the early 1900s, concrete sellers and contractors developed a variety of methods for coloring and texturing concrete flatwork. An important and popular technique was stamping the surface to create texture or relief patterns, developed in California in the 1950s. This was often combined with coloring to make flatwork look like brick or stone, or add appeal in other ways. Some referred to it as "architectural concrete."

But a very different movement arose in the San Francisco Bay area of California during the 1960s. A small number of independent artists, sculptors, and interior designers began to experiment with the use of concrete for interior fixtures and surfaces. Most were not previously involved with concrete professionally. Instead, they were attracted to the material because its properties offered great artistic potential for home interiors. Concrete's plasticity of form, mass, and uniqueness offered the potential for interiors distinctively different from what was available with the commonly used materials. Their work came to be referred to as "decorative concrete".

The decorative concrete artists developed their own methods of coloring, texturing, and shaping concrete, although they sometimes borrowed and modified the older techniques used for exterior flatwork as well. They originally applied their methods to interior surfaces, particularly counters and floors. However, over time some of their new techniques were later adopted outdoors again, and added to the range of decorative options applied to exterior flatwork. Interest in decorative concrete built gradually at first, but in the late 1990s there began a surge in its popularity. Professionals all across North America have turned to producing creative, aesthetic fixtures and surfaces with concrete. This includes artists and designers who focus on the interior, others who have expanded to working indoors, outdoors, and all around the building, and conventional concrete contractors who have learned new decorative techniques so they can offer their customers more aesthetic options.

Market

Decorative concrete has gradually moved out from the residential market. Originally, nearly all of the projects were in single-family homes. As interest and availability has grown, owners of some commercial buildings have begun to specify the material. These are mostly high-end buildings that serve customers of some type, such as retail stores and hotels. Nonetheless, probably 80 percent of decorative concrete projects today are still in homes.

Within the residential market, the interest has been shifting and spreading out. The original customers were mostly innovative individuals who enjoyed trying something different. By the 1990s interior decorative concrete had shifted to mostly upper-income homebuyers. However, today sales are also growing among the middle of the homes market.

Advantages for owners

The appeal of interior decorative concrete for owners is look and feel, plain and simple. There are sometimes other benefits, but these are almost always byproducts and not the major reason the buyer chose the product.

The techniques now used to color and texture decorative concrete produce bold and subtle appearances that cannot be exactly accomplished by any known method on any other material. Many decorative concrete projects are unique. It has become a fashion item, popular partly because it is different from anything that has been available before. The constant arrival of new techniques for texturing and coloring promises to keep it popular by keeping it changing.

Advantages for practitioners

Some decorative concrete work is designed to make it look like other materials, such as brick or stone. This has an appeal, especially where it can create the appearance of a popular product less expensively than can be done with the real thing. However, this type of work is becoming less common as "unique" concrete work gains popularity.

For contractors, the main benefit of decorative concrete is increasing the appeal of the building to the buyer. Being able to put these distinctive concrete pieces into a building may attract attention from buyers considering multiple contractors. It may also tip the balance of the final decision.

For artists and interior designers, the method offers new options for design that cannot be achieved in any other way. Unique and distinctive spaces are possible.

Decorative concrete is usually a value-added upgrade. It tends to command a premium price and give the contractor a relatively large markup that contributes to overall profit.

For some practitioners, part of the appeal of offering the product is personal. They enjoy it themselves, they enjoy making it, and they enjoy delivering it to a grateful customer.

Techniques

It is difficult to talk about the standard characteristics or parts of a decorative project. Projects are often intentionally designed to be different from one another. For the same reason, it is difficult to generalize about the installation methods used. There are, however, some favorite techniques used to add texture, pattern, or color to the concrete surface. These are combined in various ways. Although there will probably be others appearing soon after this book arrives, it is useful to understand the most popular ones.

Stamping

Stamping involves pressing a *stamp* into the surface of wet concrete to create a texture or relief pattern in it. This is something like the old practice of pressing a ring into hot wax to create a seal with a distinctive pattern to it. Modern stamps are usually a rubber-type material. They are an inch or two thick, and usually 2 feet to 4 feet on each side. They can be laid side-by-side, or picked up and reused to work across an area of concrete. The stamps are designed so they match up along the edges and create what appears to be a seamless pattern in the concrete (Figure 32-2).

Stamps can create a wide range of textures and contours in the concrete. When certain surface treatments are added afterward, these treatments may concentrate in the low spots of the stamped surface. This can create variations in the color or sheen across the concrete surface, which increases the complexity of the appearance and may enhance its appeal.

32-2 *Workers stamp flatwork.*

Stenciling

Stenciling consists of pressing a sheet of flat material into the concrete surface. The material may be plastic or a paper. It is cut into interesting shapes like the stencils used to create decorations and drawings on other surfaces. After the stencil is on, the concrete is *floated*. This means trowels or other tools are run over it to press it level and smooth. After the concrete cures the stencil is removed to reveal a shallow relief pattern (Figure 32-3).

An additional option is to spread some other treatment over the surface while the stencil is still in place and the concrete is still wet. This treatment might be some sort of coloring or texturing of the surface. When the stencil comes up this secondary treatment will be only on the areas without the stencil, sharpening the contrast of the pattern.

Pigmenting

Pigmenting involves putting pigments into the wet concrete when it is originally mixed. This is sometime called *integral coloring* because it changes the color of the material all the way through, not just on the surface. By selecting the available cement and pigments correctly, it is possible to produce virtually

32-3 *Applying stencils to decorative concrete.* Hanley Wood, LLC

any color. As noted in Chapter 3, a few will be more expensive because of the higher costs of white cement and certain pigments.

Acid staining

Acid staining occurs when solutions with certain kinds of salts are dripped onto concrete. These salts react with the calcium hydroxide of the concrete, to change its color. The results are different from other types of coloring in appealing and distinctive ways. For an artist, acid staining resembles watercolor painting, while other methods are more like oil painting (Figure 32-4).

In acid staining the color spreads unevenly, something like the way a drop of watercolor paint spreads over the surface of a wet paper. This produces delicate variations in tint and shade. It also causes different colors to bleed into one another and blend when they come into contact. The staining seeps down a bit into the concrete, changing color for a few fractions of an inch below the surface. It is slightly translucent, so viewers can see the colors at the surface and below the surface at the same time.

A growing number of contractors and suppliers are recommending precautions with the handling and disposal of acid staining materials. Some consider the acid itself toxic, and when the acid reacts with other materials in the

32-4 *Worker depositing acid stain.* Bridgeworks, Inc.

concrete it can produce metallic and other compounds. Some contractors neutralize excess acid on the floor and in their containers with baking soda, vacuum up the resulting material, and discard it as a hazardous waste. Some are moving away from acid stains altogether. New water-based alternatives are discussed in Chapter 35.

Dry shake

Powders shaken over the wet concrete can add color or other properties to the surface. This technique is called *dry shake* (Figure 32-5).

An early surface-applied powder was a *concrete hardener*. This is a material that makes the concrete on the surface become even harder than usual, making it more durable and slowing the effects of wear. It is frequently applied to concrete flatwork for that reason. However, it also affects the color of concrete slightly, making it darker. Decorative concrete contractors discovered

32-5 *Spreading dry shake.*

that it adds variation to the surface color if it is spread unevenly. This is a desirable side result. They frequently apply hardener intentionally in an uneven way to get a particular effect.

Applying a powdered pigment by the dry shake method has now become a popular coloring method. It colors the top layer of the concrete.

The usual method of application is by shaking the powder out of a tin container with holes at the top, like a flour shaker or oversized saltshaker.

Surface cutting

After the concrete has cured, workers sometimes use diamond saws to cut lines and contours into the surface. This *surface cutting* allows fine sculpting that is difficult to achieve by other methods (Figure 32-6).

Grinding and polishing

Cured concrete may be ground down with concrete grinders to create an extremely smooth surface. This may be used with almost any other treatment to create a fine, shiny floor or countertop. It is rarely used outdoors because of

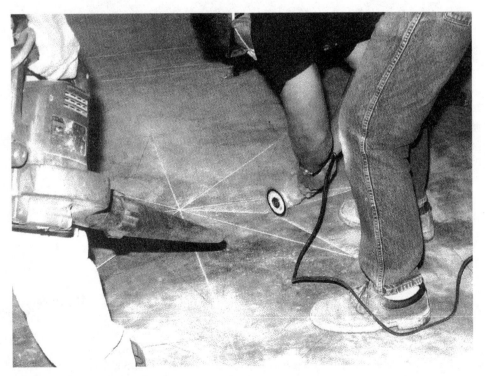

32-6 *Surface cutting of concrete.* Portland Cement Association

32-7 *The underside of a grinder used to polish a concrete floor.* Portland Cement Association

the difficulties of upkeep there. When select aggregates are mixed into the concrete, it may also expose interesting or colorful pieces of polished stone (Figure 32-7).

Embedments

A very old technique that often finds new life is *embedment* of pieces of other hard materials in the wet concrete surface to serve as a permanent decoration. The method is versatile because of the wide range of things that might be used, such as sea shells, bits of pottery, select stones, pieces of glass, and so on. They may be arranged in virtually any pattern or "randomly" to create desired effects.

Overlays

Overlays are materials applied over cured concrete in thin layers. They can create novel patterns, textures, and coloring. The materials are sometimes called *concrete overlays* or *cementitious toppings* (Figure 32-8).

Most overlays used today are a mixture of cement, some fine aggregates, and special polymer resins (which most of us would call plastics). They can be pumped or poured to produce a flat surface without troweling. They can also be treated with a trowel or other tools to create textures or patterns. Many

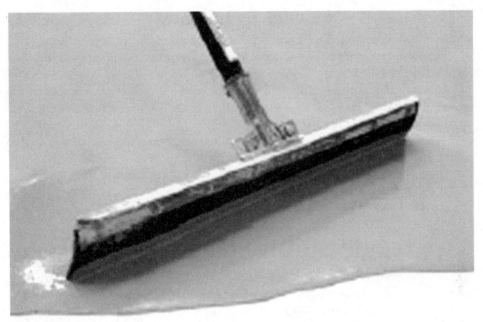

32-8 *Placing an overlay onto a floor.*

are formulated for rapid turnaround. These are ready for traffic in as little as three hours. Many are hardwearing, with strengths up to 8500 psi. They can also be applied to concrete of almost any age. This makes them useful for adding appeal to old surfaces as well as new.

It is possible to perform almost all of the treatments to an overlay that are applied to wet concrete. Some of the more popular options are:

- Stamping
- Integral coloring
- Chemical staining
- Stenciling
- Embedment
- Cutting patterns with diamond blades
- Polishing

Sealing

Although not primarily an aesthetic treatment, a common and important one is *sealing.* This is the application of a coating to the concrete that protects it from certain types of wear or discoloration. Whether any sealing products are needed and the types used depend on the other surface treatements and on the application. Sealants are therefore discussed more in the specific product chapters.

Older techniques

Some old methods used to texture the surface still pop up occasionally.

Exposed aggregate is the process of removing the top layer of cement to expose the stones in the concrete. This can be done a few different ways. Instead of a surface of one color, this produces one dotted by the colors of the stones in the mix. Particular varieties of stone may be mixed into the concrete to produce a desired final appearance.

Trowel texturing and *float texturing* involve special tools that are drawn across the wet concrete surface to create textures or surface relief. This is similar to methods of creating textured plaster on walls.

A *salt finish* is created by spreading rock salt over the concrete before floating it. The salt later dissolves away, leaving small random divots.

Code and regulatory status

Interior decorative concrete is generally not regulated. Interior floors are subject to structural code requirements. However, the surface finish is generally not related. The structural regulation of floors is covered in the chapters discussing concrete floors, which are in Part 2 of this book.

Some materials used inside a building are regulated for such properties as flammability and toxicity. However, experience and science have shown concrete to be largely safe in these regards, so that the use of exposed concrete is not of concern to officials. As a result, building officials rarely ask for information about decorative concrete or ask to inspect the work or the results.

Architects and engineers

In buildings designed by an architect, the designer is likely to be integrally involved in the planning and design of any interior decorative concrete. Larger architect-designed projects may have a separate person designated the *interior designer* who takes responsibility for the indoor fixtures and finishes. The interior designer coordinates with the architect. In that case, the designer will likely be responsible for the decorative concrete.

The involvements of the architect or interior designer may vary. Probably he or she will describe or draw the product expected, including colors, textures, and the like. The architect or designer may be involved in selecting and monitoring the artist or contractor. This is similar to what happens with any aesthetic product.

Because the finishes are unrelated to structure or mechanical matters, engineers do not get involved with them as a rule.

Maintenance and repair

Architectural concrete finishes are generally durable. However, they are likely to have different wear characteristics. These depend on the application. Details are in Chapters 34 and 35.

Sustainability

Decorative concrete finishes are generally too small a part of the materials volume of a building to be considered significant influences on its sustainability. They are rarely factored into any of the Leadership in Energy and Environmental Design (LEED) points awarded to a project.

Key considerations

Each decorative technique has things that need to be controlled to achieve good results.

In stamping, too much water in the mix can lead to water bleeding to the surface and affecting the finish. The procedure requires care and plenty of stamps on-hand to keep the alignment of the pattern accurate.

Accurately placing stenciling can be difficult when wind is blowing through the work area. This can be dealt with by starting the stenciling on a side of the formwork that is parallel to the wind direction.

With pigmented concrete, it is important to avoid certain admixtures. Accelerators, for example, may affect the final color.

Acid staining involves strong acids and can produce harmful vapors. It is important to keep all skin covered and wear goggles and a respirator. Cleanup of the surface should be done with an acid neutralizer such as ammonia or baking soda and not allowed to enter the water supply.

Dry shake depends on an even distribution of the powder for consistent coloring, if that is desired.

Grinding and cutting produce concrete dust. Workers should wear a respirator.

Successful concrete overlays depend on a properly prepared surface before the application begins. It must be clean and free of oils, sealers, and the like. It cannot be subject to high rates of moisture transmission. If these conditions are not met, the bond of the overlay to the concrete may fail.

33

Countertops

Concrete countertops bring the range of looks of decorative concrete to work and dining surfaces. They rest on conventional cabinets. The concrete material is also highly durable (Figure 33-1).

Concrete countertops may be constructed in a shop or on-site. Originally, decorative concrete artists built them in the shop, and most still favor that method because of the more specialized equipment and higher level of control available at a dedicated facility. This has made it more of a custom design and fabrication enterprise like cabinet making or sculpture and less like a typical construction trade. Prices are in the high end of the range for countertop materials. The amount of customization and individual attention is also high.

History

Decorative concrete countertops first appeared regularly in the San Francisco Bay area of California in the 1980s. They were the work of independent designers and craftsman interested in experimenting with something new. The distinctive product gained a grass roots following and momentum over the years. It spread by word-of-mouth and as the result of magazine articles and books about the interesting new technique.

The initial projects were mostly in contemporary kitchens. However, the product has spread to other rooms as well and to a much broader range of styles, including traditional.

Market

There is no precise count of the volume of concrete countertop construction. There are probably close to a thousand designer/producers for the product across North America. Almost every major metropolitan area appears to have several.

The early customers of the 1980s came from a broad spectrum of economic groups. What tied them together was an interest in things that were new and

33-1 *A concrete countertop.* Distinctive Concrete-Boston, Inc.

distinctive, rather than any particular income level. In the 1990s, the cost and high level of aesthetics of the product steered the product to the high end of the market, and the wealthy became the predominant customers.

In recent years, the market has broadened. A growing number of people are learning about the product and becoming interested. New methods have led to some lower-cost versions of the product. Chief among these is the assembly line-like production of stock countertops, instead of all custom work tailored to the individual job. This is of less interest to the wealthy, but is acceptable to some looking to spend less money. The result of all this is growing sales in the middle of the market.

Concrete countertops are still mainly a product for single-family homes. However, recently businesses have begun to take an interest because of the possibilities to achieve a distinctive interior. Some early customers are higher-end hotels, apartment and condominium buildings, retail stores, and offices.

Most of the early sales came about because homeowners heard of the technique and searched for a contractor who could do it for them. More recently, designers and contractors that offer the countertops have begun to market the product and suggest to buyers that they might want to add it. Architects have also begun to recommend it in buildings they design. This is increasingly common in commercial projects.

Advantages to the owner

Concrete countertops have become a popular aesthetic product for kitchens. Produced in antique-theme styles, they are also attractive for remodels that maintain an earlier look. They are generally considered a high-value option in interior design.

The concrete can be customized to match or contrast with the other interior features of the house, in almost any way desired. It can also be shaped to fit almost any layout without seams.

The concrete also provides a very durable functional surface.

The more unusual and intricate countertops naturally tend to cost more.

Advantages for the contractor

Striking countertops can attract buyer attention and help the designer or contractor gain a reputation as versatile and cutting edge. They typically command a significant price premium that can increase profits.

The highly custom product typically requires a close level of interaction with the customer or designer. This can strengthen the relationship and increase the likelihood of repeat business. However, it also requires time and attention.

Components

Concrete countertops are made with custom concrete mixes, reinforcement, and sometimes structural support. Each contractor has his own favorite materials to use, and there is wide variation.

In general, the concrete mixes used have small aggregate sizes. Rarely is aggregate greater than $3/8$ inches. This helps the flow of concrete into all corners of the forms to create a smooth surface. Of course, if a special stone is used to create part of the appearance of the final counter, size may be selected to be whatever is most attractive.

The water-cement ratio is typically held very low. This helps keep the strength of the concrete high, minimize shrinkage cracking, and minimize the possibility for water pooling on the form edges to create stains or voids.

Some suppliers now offer their own prepackaged concrete mixes for counters and other interior decorative concrete projects. These are particularly useful for the new user who is not ready to experiment with details of the mix yet.

In most cases reinforcement is used. This may be polypropylene fibers, glass fibers, carbon fibers, welded wire mesh, conventional rebar, or some combination of these. It further limits shrinkage cracking and helps the counter span unsupported distances. Heavier reinforcement is important in high seismic areas, which usually leads to the use of rebar. Carbon fibers are new and still somewhat expensive, but have great strength for a fiber product.

If the counter must span long distances, the contractor may use hardware to support it. This could be brackets, a skirting, or any of a dozen other items.

Counter assembly

The typical concrete counter is 2-3 inches thick. The concrete often contains fibers throughout, welded wire mesh, or rebar. Mesh or rebar is typically in one layer at about mid-depth in the concrete slab. This allows complete concrete coverage. Counters are usually connected to the cabinet below with fasteners or adhesive caulk. Fasteners used include concrete screws, brackets, or threaded rod. They are important for heavy installations and in high seismic areas.

Beyond these basic properties, there are few "typical" features to concrete countertops. Most are rectangular, but complex angles and curves and freeform shapes appear with increasing frequency. Most are flat. However, a growing number have multiple levels or waves that extend up and down. Many now incorporate one or more sink bowls in one continuous casting of concrete. Some are cantilevered out from their cabinets, just like lightweight counters.

Cost

There is no "typical" concrete countertop, so prices vary. Nonetheless, there is a surprisingly narrow price range for the custom countertops that make up most of the market. The standard price is now about $100 per square foot, including all labor, materials, installation, and contractor markup. Some producers with streamlined production techniques offer custom product for as low as $75 per square foot. And of course some highly finely produced and specialized counters can be more.

This puts concrete among the high-end materials for counters. Table 33-1 includes typical costs for the popular options.

It is difficult to give rules for how costs vary. Materials are usually a minor factor. Some techniques and combinations of techniques are more difficult than others. Irregular shapes are more difficult to produce than others. However, much of the pricing depends on buyer preferences and the style and reputation of the designer.

Some shops now produce standard counter sizes in a few styles and sell these as stock items. This is similar to selling precut slabs of stone for coun-

Table 33-1. Typical Prices of Popular Types of Countertops.

Material	Cost per Square Foot*
Stainless steel	$100–200
Concrete (custom)	$ 75–125
Granite (tile or sheet)	$ 50–100
Slate (black or red)	$ 50–100
Engineered stone	$ 50–100
Marble	$ 50–100
Hardwood	$ 50–100
Solid surface	$ 55–75
Laminate	$ 15–30

*All costs in U.S. dollars per gross square foot counter area.

ters. Because this is more efficient in production, these stock counters are sold for less. However, their artistic qualities and uniqueness are generally less.

Construction and installation

To insure a perfect fit, work begins only after the cabinets are installed permanently in place. The contractor takes exact dimensions at the site. There are usually consultations with the client to determine the thickness, shape, colors, finishes, and features desired.

The artist or contractor prepares small samples of the planned material and submits these to the customer for approval. Typically these are saved to check against the final product.

Shop production

Work to create a counter in a shop begins with construction of a form. The contractor uses the measurements to create a template of the counter shape in lightweight sheet material. Once this is checked, it is used to transfer the outline to the form material.

Counters are typically cast top-down, and the form is constructed accordingly. The form is usually a smooth material. It needs to be stable and nonabsorbent. Commonly used are melamine-coated fiberboard, metal sheet, and plastics. Blockouts are added to create openings for such things as sinks, drains, and faucets. Some suppliers now offer preformed blockouts in shapes and sizes that match standard plumbing fixtures. These are usually rubber for easy removal and reuse. The sidewalls are attached around the perimeter. They must be supported with backer boards so they do not deform under the pressure of the concrete. All seams must be sealed (Figure 33-2).

33-2 *Placing concrete at a shop.* Stone Soup Concrete

Any mesh or rebar that is installed must stop short of any edge by at least 1½ inches.

The concrete mix is highly individual to the contractor and the job. Mixing it usually involves a high level of care in getting exact proportions and timing. It is normally placed by shovel or scoop. It is consolidated thoroughly with a small vibrator.

After 20 minutes to 30 minutes the contractor trowels the upper surface smooth.

After the concrete cures, the form comes off. The concrete may be finished with cleaning, grinding, polishing, or adding inlays. Many contractors also add a sealant to the surface.

Installation must be onto cabinets of adequate strength. Some production cabinets cannot support the weight. Typically cabinet walls of at least ¾-inch plywood are adequate for counters up to 2 inches thick. The contractor may need to modify or add to the cabinets. When changes are needed, most often it is the toe kicks and bases that need to be replaced with a stouter wood structure.

The top of the cabinets should be covered with a single solid layer of plywood to support the weight evenly.

The contractor next sets the counter in place to check for fit to the cabinets and any leveling. The plywood substrate must be cut for the openings.

When all is ready, the contractor masks the cabinets as needed and puts a layer of adhesive caulk over the top of the plywood substrate. The counter goes over this, set in precise position. It remains undisturbed for a day until the adhesive cures (Figure 33-3).

In some cases the contractor also drills through the bottom of the plywood substrate and into the concrete to secure the counter further with concrete fasteners. These are especially important for heavy installations and in high seismic areas.

33-3 *Installing the final product.* Stone Soup Concrete

In-place production

Some countertops are created on the cabinets at the project site instead of in a shop. This eliminates the need for a shop or moving the final product. However, it requires some heavy temporary support for the forms and can be messy.

When forming the countertop in-place, it is important to cover all surfaces underneath to catch loose concrete. If the countertop will be backed by gypsum wallboard on one side, the wallboard should be painted there to prevent it from absorbing water.

Cement board may be used for the bottom support instead of plywood because it will not warp when wet. Temporary plywood supports brace it from below. These can also form a "lip" in the counter over the edge of the cement board. Plywood edge pieces form the sides of the counter. The plywood forms are sealed and the joints caulked. Knockouts for sinks and other fixtures need to be taped so that they are easily removed.

Reinforcement goes in and concrete is placed and consolidated as usual. The top surface is typically floated every few minutes for several hours to produce the finest possible surface. The edge forms may come off partway through this process to allow finishing of the sides.

Connections

As noted in previous sections, concrete countertops are secured to a plywood substrate with adhesive caulk. Sometimes they are further secured with concrete fasteners through the substrate and into the bottom of the concrete.

Any plumbing is connected through appropriate holes in the concrete as it is to any common countertop.

Architects

Architects and interior designers are increasingly involved with the concrete countertops. This is especially true in large projects, which are traditionally designed by architects.

Frequently it is the architect or designer that recommends the use of concrete counters to the client. These professionals increasingly have their favorite styles and contractors to work with. Early in the process, they may describe or sketch some guidelines as to the style of counters they are looking for. Typically, the samples are submitted to them for approval.

The actual production and installation of the counter tends to fall late in the construction of a project. At that point there may be little interaction with the architect or designer once design has been approved.

Training

There is no required training for a countertop producer. However, a beginner can benefit from the experience of others by taking a class or seminar on the subject. These are offered at some concrete trade shows and by some of the longer-established decorative concrete producers who have created more substantial organizations. Two of these are the Concrete Countertop Institute (www.concretecountertopinstitute.com) and the Cheng Concrete Exchange (www.chengconcreteexchange.com). It is sometimes possible to find other classes by contacting nearby decorative concrete artists/contractors. They may offer such training, or know of others that do.

Maintenance and repair

Concrete countertops will remain functional for decades without any significant maintenance. However, if the counter goes unattended, under the wear of knives and plates and pots it may lose some of its luster and develop some variations in color. Some owners find this attractive but others do not.

To prevent changes in appearance, most producers recommend sealing the surface with beeswax or a commercial sealer periodically. Recommendations for the frequency of sealing vary from months to years. Best is to consult the producer for advice.

Day-to-day the owner can limit wear by limiting acidic fluids that might spill onto the counter and avoiding use of abrasive pads or cleaners.

Some producers offer to refinish their counters to return them to their original appearance. This may be called for after 5–8 years.

Key considerations

Because of differences in materials and conditions, it is not always possible to get exactly consistent appearance from one counter to the next or evenly across a single counter. This natural variation is usually considered attractive. However, both the producer and the customer should be aware of it and willing to accept it.

Prolonged contact of the skin to wet concrete can cause irritation and burning. It is standard recommended procedure to wear gloves, goggles, and long sleeves and pant legs when working with concrete.

Any sealants used on the surface should be certified as food safe.

Getting a fine, smooth, consistent concrete surface requires care in the forming and placement. The forms must be carefully sealed, nonabsorbent, and free of leaks or debris.

Availability

There are now contractors that regularly produce concrete countertops throughout North America. Unfortunately, finding them can be difficult because they are mostly small and there is no separate listing for them in most directories. Directories specifically for concrete countertop producers can be found on the Portland Cement Association web site www.concretehomes.com (Click on "Building Systems" and then on "Countertops") and the independent web site www.concretenetwork.com (Click on "Concrete Countertops" and then on "Find a Countertop Manufacturer/Designer").

The Concrete Countertop Institute (www.concretecountertopinstitute.com) and the Cheng Concrete Exchange (www.chengconcreteexchange.com) now certify designer/installers according to the courses they have taken and their work experience.

Support

The organizations that do training will sometimes field requests for information and provide publications. The Portland Cement Association also offers publications on its web site, www.concretehomes.com. Information on countertop construction and some of the aesthetic treatments is also available on the independent web site www.concretenetwork.com.

Current projects

The best bet to locate current projects for viewing is to contact local producers.

34

Decorative floors and flatwork

The concrete slabs we walk on can be treated with decorative concrete techniques to produce striking aesthetic finishes. In recent years the creation of decorative flatwork has grown into a regular trade that is now widely available. Building owners are buying this service in rapidly growing numbers. Designers and contractors are realizing solid profits for a product that is in high demand (Figure 34-1).

Technically, the term *flatwork* includes any horizontal concrete slab cast on the ground or formwork. It therefore includes indoor floors as well as outdoor drives and walks. To make it clear which form of flatwork they are discussing, people often refer to indoor slabs as "floors" and outdoor slabs as "exterior flatwork."

Virtually all the common decorative concrete treatments can be installed on a flatwork surface.

History

Decorating concrete floors goes back to the very first uses of the material in ancient civilizations. The Romans and Greeks arranged tiles in their floors. Even earlier peoples applied paint, pigments, or indentations to them.

Particularly important in the early development of architectural flatwork was a contractor named Brad Bowman. He developed new stamping techniques in Southern California in the 1950s. These methods achieved new levels of detail, patterning, and practicality compared with old ways to indent wet concrete. Methods for coloring the concrete, some of them developed earlier, were added to the textured flatwork. Over time, these new techniques spread to Northern California and across North America. They were practiced mostly by traditional concrete contractors.

Interior decorative floors of the type popular today first appeared in the San Francisco Bay area of California in the 1980s. This was part of the general

34-1 *A decorative concrete floor.* Distinctive Concrete-Boston Inc.

development of modern decorative concrete. The producers were independent designers and artists experimenting with techniques to produce new and unique finishes. Although some specialized, others produced both floors and counters, and other interior surfaces and fixtures as well.

They went far beyond Bowman's techniques, however. They adapted or developed many others methods of coloring, texturing, and patterning the concrete surface in what has become a steady wave of art and experimentation. Many of the techniques they developed indoors later found their way outside to bring these new options back to exterior flatwork constructed by concrete contractors.

As with the countertops, word got out about the new floors and flatwork, and their popularity spread. Today, there are probably about a thousand contractors regularly doing decorative treatments of concrete slabs, spread across North America.

Market

The market for decorative flatwork is very similar to that for concrete countertops, except that the flatwork has gone even further in penetrating the non-residential market.

There is no precise tally of the amount of decorative flatwork production in the United States or Canada, but there appear to be multiple contractors and craftspeople doing the work regularly in almost every metropolitan area.

The original customers of the 1980s included artistic and curious homeowners. After the earliest jobs received publicity, interested owners began to seek out the designers to do this work for them. The product began to sell mostly to the high end of the home market by the 1990s. At this time, growth also accelerated.

In the last few years, sales have broadened to include more homeowners in the middle of the market. In addition, businesses have commissioned decorative floors and exterior flatwork in rapidly growing numbers. They consider decorative concrete floors and walks to be an effective way to attract or impress discriminating customers and tenants. As one might expect, the sales are concentrated in higher-end hotels, apartment and condominium buildings, retail stores, and office buildings.

Current sales are based less on requests from buyers than they once were. The artists and contractors have begun to market their services actively. Larger building projects usually involve an architect or interior designer, and these professionals are increasingly aware of decorative concrete floors. They often specify the floors or recommend them to the client.

Advantages to the owner

Decorative concrete flatwork has become a fashion and status product for homes. In commercial establishments it contributes to the desired atmosphere.

The concrete can be tailored to achieve a wide range of desirable appearances. It can be closely coordinated with the rest of the décor.

The concrete surface is durable. In some stores it has been made to resemble softer materials such as wood planks. Decorative concrete provides the desired appearance, but with less wear or maintenance

Depending on the treatments chosen, decorative floors and exterior flatwork generally carry a premium price tag. They must also be custom produced on-site.

Advantages for the practitioner

Decorative concrete attracts buyers. The exterior flatwork has curb appeal. The floors can be elegant and striking. They fetch premium prices that can support good profit margins.

Producing them requires working closely with the client. This builds an important relationship, but also requires time and attention.

Flatwork assembly

The decorative work is usually a surface treatment. It may be applied to almost any type of concrete slab, which is constructed by the usual methods. Generally speaking, the decorative finish has little influence on the basic flatwork construction and vice-versa.

Cost

Subcontractors typically charge $5.00 to $6.00 per square foot for a basic decorative concrete floor or flatwork. This includes the concrete material and the labor to place it to create the flatwork, in addition to the decorative treatments. It also includes the subcontractor's markup.

This compares to about $2.00 to $3.00 for a plain concrete floor or slab on grade. It is more than carpet or vinyl flooring, but usually less than a tile, natural stone, or hardwood floor.

Typical materials and labor costs for a simple colored and stamped floor are summarized in Table 34-1. These do not include a subcontractor markup.

Costs may be lower for larger installations. The workers typically gain efficiency over a larger area.

Costs can vary sharply depending on exactly what treatments and what level of detail and finish are required. Especially popular in interior floors are complex applications of staining and texturing that can add several dollars per square foot to costs.

Installation

Some decorative techniques are performed in the wet concrete immediately after it is cast. These include stamping, stenciling, embedments, and dry shake. Others are performed on cured concrete. These include acid staining, diamond cutting, and overlays.

Table 34-1. Typical Costs for a Basic Decorative Concrete Floor.

Item		Cost per Square Foot*
Labor		$3.00
Materials		$1.45
Concrete	$1.00	
Reinforcement	$0.25	
Pigments	$0.20	
Total		$4.25

*All costs in U.S. dollars per gross square foot of surface area.

The time of installation makes important differences for certain types of flatwork. With exterior flatwork, work in wet concrete is common. The decorative crew usually also lays the flatwork. The crew is fully prepared for both tasks in advance, and the workers can begin applying the decorative treatments as soon as they level the top surface. If the project requires treatments on the cured concrete, they come back to do that later.

With interior floors it is not always as easy to perform treatments in a wet concrete slab. The slab is typically cast by a different crew. The decorative crew would have to coordinate closely with them and be on-site as the slab work ended to begin the decorative work. The finished floor surface would have to be covered and protected while the rest of construction proceeded. This can be done, but it requires more planning.

More often, decorative floors are created in other ways. When the desired techniques are those performed in cured concrete, the decorative work simply comes later. Floor covering is typically the last major task of building construction, anyway. The decorative work is simply performed at that point. Some basic protection for the plain concrete surface before then is advisable.

If the floor treatments must be made in wet concrete, the contractor may pour a thin additional layer of concrete over the surface and put the treatment on that. This is also toward the end of building construction. It has the additional advantage of allowing the decorative contractor to use his own, controlled mix of concrete.

After the treatments are finished and the concrete is fully cured, the contractor usually applies a protective sealer. The floor is also covered for the rest of construction, as with any fine flooring.

Architects

In buildings employing professional designers, the architect or interior designer is usually closely involved in the decorative flooring. The building architect or a landscape architect may likewise be involved with any decorative exterior flatwork. In many or most cases, it is the architect or designer who recommends these products to the client. Increasingly, they have favorite styles and contractors they prefer to work with. They will give design guidelines or even detailed sketches to the artist/contractor to show what they want the final floor to look like.

The contractor usually prepares and submits samples of treated concrete for approval. Especially in large and high profile projects the architect or designer may even require construction of a mockup slab on-site. In some cases the architect or designer may wish to be present for the actual work to observe the application of materials as specified.

Training

There is no required training for a decorative flatwork contractor. However, the skill involved is high, and the beginning contractor may wish to get instruction. Some of the larger suppliers of pigments, stamping equipment, and other accessories offer periodic training workshops at their regional supply centers. Contact information for some suppliers is available on the Portland Cement Association web site, www.concretehomes.com. Click on "Building Systems," then on "Decorative Flatwork." The listing of providers includes contractors and some suppliers.

Maintenance and repair

Concrete is a highly durable form of floor or walkway. The decorative treatments can wear over the years as a result of traffic. This wear may be considered attractive, similar to antiquing effects. However, some owners may wish to prevent it. In this case, a sealer should be applied to the surface periodically.

Penetrating sealers form a hard resin chemically bonded to the concrete surface. They are solvent-based with fumes that should not be inhaled in any significant amount. For that reason they are often restricted from installation in high-traffic public areas.

Film-forming sealers create an adhesive sealing barrier on the surface of the concrete, filling holes and forming a coating to protect from water and chemicals. These are often water-based, with little odor or potential for toxicity. They can be worn off over time and should be reapplied periodically. They require some drying time after application before they can take traffic.

Usually the artist/contractor who did the work can recommend a specific sealer.

Key considerations

Efficient production of quality decorative concrete requires coordination between the placement of the concrete and application of the surface treatments. When this involves two different crews, they need to discuss their work in advance and agree on exactly what concrete mix will be used.

There are many considerations involved in using one or another of the available decorative techniques. These are discussed in Chapter 32.

Availability

Contractors doing decorative slab work are available throughout North America. They tend to be small operations and are not included as their own category in most directories. A directory specifically for decorative floor contractors is on the Portland Cement Association web site www.concretehomes.com. Click on "Building Systems" and then on "Decorative Flatwork." Other directories that include decorative flatwork contractors are on the independent web site www.concretenetwork.com. Click on "Concrete Patios," "Interior Concrete Floors," "Decorative Concrete," or "Concrete Driveways." Then click on "Find a Contractor."

Support

The large suppliers of pigments, stamps, and related supplies have technical staffs that can answer many questions. The Portland Cement Association also offers publications on its web site www.concretehomes.com. Introductory information and some publications are also on the independent web site www.concretenetwork.com.

Current projects

The best bet to locate current projects for viewing is to contact local contractors.

35

Developments in decorative concrete

The field of decorative concrete is constantly seeing new finish techniques. Because its purpose is aesthetic, it is largely a "fashion" trade. The craftspeople and customers alike are constantly seeking new and different techniques that produce appearances not previously available and bring the practice into new applications.

New applications

From the beginnings of decorative concrete in the 1980s, interior designers were interested in taking advantage of the capabilities of decorative concrete in more than just countertops and floors. As interest has grown, they have increasingly been taking their art to other areas. Inside of buildings they are building whole other fixtures and surfaces with concrete so that they can get its design properties in new and interesting places (Figure 35-1).

Some of the other applications are wall dividers, fountain works, showers, built-in bathtubs, shelving, tables, and other fixtures and even furniture.

The story is similar outdoors. Decorative techniques once used primarily for walks and drives are finding their way onto all sorts of flatwork. This includes patios, pool decks, and more.

Together these new applications add up to a trend, which is experimenting with the possible uses of decorative concrete in and around a building. Exactly where it ends up is difficult to say, but where it is heading is clear. It is leading us to greater variety and a higher degree of design throughout our buildings and grounds.

Water-based stains

For decades now, concrete stains in a water base have been available. However, they have not always been as flexible and useful as acid-based

35-1 *Bathroom with concrete floor, wall surfaces, and fixtures.* Cheng Concrete Exchange and Taunton Press

stains. The acid-based products have become more widely used for decorative concrete.

More recently, sellers have developed vastly improved water-based stains. Like acid stains, they can be applied to concrete after it has cured to impart a color that penetrates below the surface. However, they work differently (Figure 35-2).

Acid stains react chemically with ingredients of the concrete, changing the concrete's color. The reaction has some natural variation. This causes the coloring to vary delicately in the concrete, much like thin water color on paper. Water-based stains contain micro-sized particles that seep into the natural pores of concrete. Depending on how they are applied, they can tint the concrete or create nearly opaque color, more like an oil paint on canvas.

Water-based stains have unique characteristics that can be advantageous. They have a wider possible color range than acid stains. In fact, virtually any color is possible. They can make brighter colors. They do not involve use of acids, which must be handled, neutralized, and disposed of carefully. It is pos-

35-2 *Worker applying water-based stain.* Hanley-Wood, LLC

sible decades later to re-apply water-based stain and return the surface to nearly its original appearance.

Water-based stains are adding a valuable new tool to the decorative concrete kit. They can be used alone or combined with other decorative techniques.

Conclusion

The dramatic growth in the use of concrete construction products will continue. It is not a fad. The forces behind it are long-term trends that will outlive monthly fluctuations in material prices and this year's design tastes. Almost any builder who expects to be operating for another ten years cannot hide from these products. Learning more about them will be a requirement of the business.

On the one hand, buyers' incomes are rising. As a result, they are gradually coming to spend more and more on their buildings. They are expecting that they will be getting more and more from them. It is true that they will devote part of their income to having more floor space constructed by the same old methods. But some income also goes to getting more out of the space they have. That is why over the years buildings have gradually came to incorporate such things as indoor plumbing, electrical wiring, wind and seismic resistance, and energy efficiency.

Concrete has properties that buyers want. These are things like durability, disaster resistance, low maintenance, and thermal mass and air tightness for comfort and energy efficiency. In the past, they may not have always been able to afford as much of these things as concrete offers. But with rising incomes, they can afford them more and more. The results are simple: People buy more concrete products.

On the other hand, concrete offers buyers more and more. New pigments and finishing methods provide premium looks that delight them and add distinction to their buildings and grounds. New product designs and admixtures give the material new flexibility and streamline installation. This makes new uses possible and reduces costs. In more and more areas, concrete may still be the premium option, but it is no longer always the expensive option. Concrete products and systems are cost-competitive with the alternatives. In some cases they are now among the *least* expensive options.

Who could turn down a product like this? If it has things you've always wanted, it was just improved to have even more, and it is about the same price as what you used to buy, switching is a natural choice. More and more concrete products meet this description. It is no surprise that more and more people are buying them.

As a contractor, you need to learn more about these products to decide whether, when, and how to adopt them yourself. For some of you, these products will be an opportunity. They are an opportunity to provide something different and better to buyers, to stand out from the crowd, and to attract more business and premium prices. But even for contractors who are not inclined to change, it may not be possible to sit back and watch as the rest of the industry starts offering these products. At some point people will expect them of you, and sales can suffer if you don't have some understanding of them.

Change can be difficult, but it can be exciting. We hope this book has helped to ease the difficulty and increase the excitement.

Index

About the authors

Peiter A. VanderWerf, Ph.D., has been researching concrete construction systems for eleven years, with an emphasis on systems used to construct homes and small buildings. He is lead author on the original *Concrete Homebuilding Systems*, and three later books on insulating concrete forms. He has written several articles on concrete wall and floor systems, and writes a regular column for *Permanent Buildings and Foundations*, a trade periodical covering small concrete buildings. He is the designer of several concrete products and systems for small buildings and consults regularly to manufacturers in the field. He received his Ph.D. in management of technological innovation from the Massachusetts Institute of Technology.

Ivan S. Panushey has researched concrete building systems and materials for six years. He has learned details of concrete construction first-hand by laboring on job sites. Among other projects, he headed a year-long study to establish guidelines for connections of steel subassemblies to concrete structural walls in small buildings. He is a member of the American Society of Civil Engineers and the National Society of Professional Engineers. He holds a bachelors degree in mechanical engineering and materials science and a masters degree in design technology and management, both from Harvard University.

Mark Nicholson is an experienced finish carpenter, cabinet and furniture maker, and ceramics designer. He has worked for several years on conventional construction projects. He has recently attended formal training in the construction of concrete homes and worked on a crew that built two houses. He holds a bachelor of fine arts degree from Alfred University.

Daniel Kokonowski has researched a wide range of concrete products and systems for over two years. He has received training and worked on crews for tilt-up, concrete masonry, and various other types of building construction. He is an accomplished graphic designer and is currently enrolled in the architecture program at the Boston Architectural Center.